Modular Arithmetic

N.B. Singh

DEDICATION

To Nature,

I dedicate this book to you, the source of all life. You are my inspiration, my teacher, and my friend.

Thank you for teaching me about the beauty of the world around me. Thank you for showing me the power of the natural world. Thank you for giving me a sense of peace and tranquillity.

I promise to do my part to protect you and your many wonders. I will teach my children about the importance of conservation and sustainability. I will work to make the world a better place for all living things.

Thank you for everything, Nature.

With love,

N.B Singh

Contents

7   Modular Linear Equations   77

Welcome to the world of "Modular Arithmetic." This book is a comprehensive exploration of the fundamental concepts, theorems, and applications of modular arithmetic, a branch of number theory with far-reaching implications in various fields of mathematics and beyond.

Purpose of the Book

The primary purpose of this book is to provide readers with a clear and structured understanding of modular arithmetic, starting from its basic principles and progressing to advanced topics. Whether you are a student seeking a solid foundation, a researcher delving into specialized areas, or a curious mind eager to explore the beauty of modular arithmetic, this book aims to cater to a diverse audience.

Organization of the Book

The book is organized into several chapters, each focusing on specific aspects of modular arithmetic. We begin with the foundational concepts, including congruence, residue systems, and modular operations. Subsequent chapters delve into applications in cryptography, coding theory, number theory, and advanced topics such as Fermat's Little Theorem and modular forms.

Features of the Book

- **Clear Explanations:** Each concept is explained in a clear and accessible manner, with step-by-step derivations and examples to aid understanding.

- **Examples and Applications:** Throughout the book, we provide numerous examples and real-world applications to demonstrate the relevance and practical utility of modular arithmetic.

- **Exercises:** To reinforce learning, each chapter includes examples.

- **Advanced Topics:** For those interested in deeper explorations, the book covers advanced topics such as the Chinese Remainder Theorem, Fermat's Little Theorem, and their applications in diverse fields.

Who Should Read This Book

This book is designed for a wide audience, including undergraduate and graduate students studying mathematics, computer science, or related disciplines. Researchers, educators, and professionals seeking a comprehensive reference on modular arithmetic will also find this book valuable.

How to Use This Book

For optimal learning, we recommend reading the chapters sequentially, as each builds upon the concepts introduced earlier. The exercises at the end of each chapter serve as an opportunity to reinforce your understanding and gain practical experience.

Final Note

We hope this book inspires curiosity and a deeper appreciation for the elegance and versatility of modular arithmetic. Whether you are embarking on a course of study or simply exploring out of curiosity, may this journey through modular arithmetic be both enriching and enjoyable.

Chapter 1

Introduction

1.1 Background

Modular arithmetic is a fundamental branch of number theory that deals with the arithmetic of integers under the operation of modulo, often denoted by the symbol %. The concept has ancient roots, finding applications in various fields such as cryptography, computer science, and number theory. At its core, modular arithmetic explores the remainders of division, offering a unique perspective on the properties and patterns within the set of integers.

1.1.1 Definition of Modular Arithmetic

In modular arithmetic, two integers a and b are said to be congruent modulo m (denoted as $a \equiv b \pmod{m}$) if they have the same remainder when divided by m. The modulo operation is defined as follows:

$$a \equiv b \pmod{m} \iff (a - b) \text{ is divisible by } m$$

For example, consider $a = 15$ and $b = 3$ with modulo $m = 4$. We can express this congruence as $15 \equiv 3 \pmod{4}$ because $(15 - 3)$ is divisible by 4.

1

1.1.2 Modulo Operations

Modular arithmetic involves basic operations like addition, subtraction, multiplication, and division. These operations exhibit interesting properties that distinguish modular arithmetic from regular arithmetic. Let's explore some examples:

Addition and Subtraction

The sum and difference in modular arithmetic can be computed by performing the operation in the regular sense and then taking the result modulo m. For instance:

$$(7 + 5) \bmod 3 = 12 \bmod 3 = 0$$

Similarly, subtraction works the same way:

$$(10 - 8) \bmod 5 = 2 \bmod 5 = 2$$

Multiplication

Multiplication in modular arithmetic involves multiplying the numbers and then taking the result modulo m:

$$(6 \times 4) \bmod 7 = 24 \bmod 7 = 3$$

Division

Division is a bit more involved and requires the concept of modular inverses. Given a, b, and m, the modular inverse of a modulo m is an integer x such that $(a \cdot x) \bmod m = 1$. Division can then be computed as multiplication by the modular inverse:

$$\frac{12}{3} \bmod 5 = 12 \cdot 3^{-1} \bmod 5$$

This brief overview lays the foundation for exploring the rich world of modular arithmetic. In the subsequent sections, we will delve deeper into its applications, properties, and advanced topics.

1.2 Motivation

The motivation behind the study of modular arithmetic lies in its versatility and its practical applications in various domains. One of the key motivations is its role in solving problems related to remainders and cycles, which arise frequently in real-world scenarios.

1.2.1 Solving Remainder Problems

Consider the simple problem of distributing 27 apples equally among 4 friends. In standard arithmetic, we would say each friend gets $27\nabla \cdot 4 = 6.75$ apples. However, in practical terms, we can't give a fraction of an apple. This is where modular arithmetic becomes valuable. We can express this situation using the modulo operation:

$$27 \equiv 3 \pmod{4}$$

This congruence tells us that after giving each friend 6 apples, there are 3 apples left, forming a cycle. Modular arithmetic enables us to precisely describe and work with such situations.

1.2.2 Clock Arithmetic

Another motivating example is the representation of time on a clock. A clock has 12 hours, and after reaching 12, it resets to 1. If it's currently 8 o'clock and we want to know the time after 9 hours, we can use modular arithmetic:

$$(8 + 9) \bmod 12 = 5$$

So, the time would be 5 o'clock. This cyclic behavior is a fundamental aspect of modular arithmetic and is evident in various cyclic phenomena in nature and man-made systems.

1.2.3   Error Detection in Computer Science

In computer science, modular arithmetic is employed in error detection and correction codes. When transmitting data, errors may occur. Modular arithmetic helps detect errors by introducing redundancy in the data. For instance, the widely used checksum involves summing up all the data values and taking the result modulo a prime number. If the received data doesn't match the checksum, an error is detected.

1.2.4   Public Key Cryptography

Motivation for studying modular arithmetic extends to the realm of cryptography, especially in the development of public key cryptosystems. Algorithms like RSA rely heavily on modular arithmetic properties. The security of these systems is grounded in the difficulty of factoring large composite numbers, which is a problem well-suited for modular arithmetic techniques.

1.2.5   Efficient Data Storage

In computer memory and data storage, modular arithmetic aids in efficient addressing and indexing. Memory addresses often wrap around in a cyclic manner, and modular arithmetic is employed to manage this cyclic behavior, optimizing memory usage.

1.2.6   Puzzle Solving

The study of modular arithmetic is also motivated by its application in solving various puzzles and games. Modular arithmetic provides a systematic way of analyzing and solving problems involving cycles, remainders, and patterns.

1.2.7 Congruences in Nature

Nature exhibits numerous examples of congruences and cycles. The study of seasons, lunar phases, and other natural phenomena often involves modular arithmetic. The periodicity observed in nature can be accurately described and predicted using modular arithmetic concepts.

1.2.8 Resilience to Integer Overflow

In computer programming, particularly in low-level languages, integer overflow can occur when the result of an arithmetic operation exceeds the maximum representable value. Modular arithmetic can provide a form of resilience by ensuring that the result wraps around within a specified range, preventing overflow-related issues.

1.2.9 Enhanced Algorithmic Efficiency

Algorithms in various domains, such as optimization, graph theory, and numerical methods, leverage modular arithmetic for enhanced efficiency. Modular arithmetic often simplifies complex mathematical expressions, leading to more elegant and computationally efficient algorithms.

1.2.10 Historical Significance

The historical significance of modular arithmetic is another motivation for its study. Ancient civilizations, such as the Chinese and Egyptians, used modular arithmetic in solving practical problems, including calendar calculations and agricultural planning. Understanding the historical roots provides a deeper appreciation for the enduring relevance of modular arithmetic.

1.2.11 Coding Theory Applications

In communication systems, modular arithmetic finds applications in coding theory. Error-correcting codes, like Reed-Solomon codes, rely on the principles of modular arithmetic to ensure the reliable transmission of information in the presence of errors.

1.2.12 Parallel Processing and Cryptocurrency

With the advent of parallel processing and the rise of cryptocurrencies like Bitcoin, modular arithmetic has gained prominence. Cryptographic hash functions and proof-of-work algorithms, essential components of blockchain technology, heavily rely on modular arithmetic.

1.2.13 Abstract Algebra Connection

Motivation for studying modular arithmetic extends to its connections with abstract algebra. Modular arithmetic introduces students to algebraic structures such as rings and fields, laying the groundwork for more advanced algebraic concepts.

1.2.14 Bridge to Advanced Mathematics

Lastly, modular arithmetic serves as a bridge to advanced mathematical topics. The study of modular forms, elliptic curves, and algebraic geometry often builds upon the foundational concepts of modular arithmetic, making it a crucial stepping stone for students pursuing higher-level mathematics.

In conclusion, the motivation for delving into modular arithmetic is multifaceted, encompassing practical problem-solving, applications in diverse fields, historical significance, and its role as a gateway to advanced mathematical theories. The richness and versatility of modular arithmetic make it a captivating and essential subject for exploration and understanding.

1.3 Scope and Objectives

In this section, we delineate the scope and objectives of our exploration into modular arithmetic, outlining the key areas of focus and the goals we aim to achieve. Modular arithmetic offers a broad landscape with applications ranging from elementary problem-solving to advanced cryptographic systems, making it essential to define the boundaries of our study.

1.3.1 Scope of the Study

The scope of our exploration encompasses a comprehensive understanding of modular arithmetic, from its foundational principles to its advanced applications in various disciplines. We will delve into the intricacies of basic modular operations, properties, and their mathematical implications. Additionally, we will explore the applications of modular arithmetic in fields such as cryptography, coding theory, and computer science.

1.3.2 Foundational Concepts

Our exploration begins with an in-depth examination of foundational concepts, including congruence, modulo operations, and the basic arithmetic operations in the modular setting. We aim to provide a solid mathematical groundwork, ensuring a clear understanding of the fundamental principles that underpin modular arithmetic.

1.3.3 Properties and Patterns

Understanding the properties and patterns inherent in modular arithmetic is a key objective of our study. We will explore the properties of congruence, the distributive property, and other algebraic characteristics specific to modular arithmetic. Through illustrative examples, we aim to elucidate the unique behavior and structure of modular arithmetic.

1.3.4 Applications in Cryptography

One of our primary objectives is to elucidate the role of modular arithmetic in cryptography. We will explore its applications in public key cryptosystems, emphasizing the significance of modular arithmetic in ensuring the security and confidentiality of digital communication. Examples will be provided to illustrate the cryptographic algorithms that leverage modular arithmetic.

1.3.5   Coding Theory and Error Correction

Another significant area of focus is the application of modular arithmetic in coding theory. We will investigate how modular arithmetic is utilized in the design of error-detecting and error-correcting codes. Examples will demonstrate the effectiveness of modular arithmetic in enhancing the reliability of data transmission and storage.

1.3.6   Advanced Topics and Theoretical Foundations

Our study extends to advanced topics, including the Chinese Remainder Theorem, Fermat's Little Theorem, and finite fields. We aim to provide a bridge to more advanced mathematical theories, offering readers a pathway to explore abstract algebra and its connections with modular arithmetic.

1.3.7   Practical Problem-Solving

Throughout our exploration, we will emphasize the practical aspects of modular arithmetic in solving real-world problems. Examples will be drawn from diverse domains, showcasing how modular arithmetic can be applied to address challenges in areas such as logistics, scheduling, and resource allocation.

1.3.8   Algorithmic Efficiency

Understanding the efficiency of algorithms is a critical objective. We will investigate how modular arithmetic can contribute to algorithmic efficiency in various computational tasks. This includes exploring algorithms that leverage modular arithmetic to optimize computations and streamline processes.

1.3.9   Interdisciplinary Connections

Our study will highlight the interdisciplinary nature of modular arithmetic, showcasing its connections with computer science, mathematics, and engineering. By exploring these connections, we aim to provide readers with a holistic view of the diverse applications and implications of modular arithmetic in different disciplines.

1.3.10   Educational Impact

An overarching objective of this exploration is to elucidate the educational impact of modular arithmetic. We aim to present the material in a manner that is accessible and engaging for a broad audience, from students new to the subject to researchers seeking advanced insights. Our goal is to contribute to the educational landscape by fostering a deeper appreciation for the beauty and utility of modular arithmetic.

1.3.11   Research Opportunities

Identifying research opportunities is a crucial aspect of our study. We will highlight open problems, current research trends, and potential areas for exploration within the realm of modular arithmetic. By doing so, we aim to inspire further research and contribute to the ongoing development of this vibrant field.

1.3.12   Pedagogical Resources

To support our educational objectives, we will develop pedagogical resources, including examples, exercises, and illustrations. These resources are designed to enhance the learning experience and provide readers with practical tools for mastering the concepts presented in this book.

1.3.13   Practical Implementations

Our study will not only explore the theoretical foundations of modular arithmetic but also delve into practical implementations. This includes examining how modular arithmetic is implemented in computer systems, cryptographic protocols, and coding practices. Real-world examples will illustrate the tangible impact of modular arithmetic in various applications.

1.3.14   Challenges and Limitations

Acknowledging the challenges and limitations of modular arithmetic is an integral part of our study. We will discuss scenarios where the assumptions of modular arithmetic may not hold and explore potential pitfalls in its application. By addressing these aspects, we aim to provide a nuanced understanding of the scope and constraints of modular arithmetic.

Chapter 2

Basic Concepts

2.1 Definition of Modular Arithmetic

Modular arithmetic is a branch of number theory that deals with the arithmetic
of integers under the operation of modulo. The essence of modular arithmetic
lies in the concept of congruence, which provides a systematic way of dealing
with remainders.

2.1.1 Congruence Relation

In modular arithmetic, two integers a and b are said to be congruent modulo m
if they have the same remainder when divided by m. This is denoted as $a \equiv b$
$(\mod m)$, and it can be expressed as:

$$a \equiv b \pmod{m} \iff (a - b) \text{ is divisible by } m$$

For example, consider $a = 17$, $b = 5$, and $m = 6$. We can say $17 \equiv 5 \pmod{6}$
because $17 - 5 = 12$ is divisible by 6.

2.1.2   Modulo Operation

The modulo operation, often denoted as $a \bmod m$, returns the remainder when a is divided by m. This operation plays a central role in modular arithmetic:

$$a \bmod m = a - m \left\lfloor \frac{a}{m} \right\rfloor$$

For instance, 13 mod 4 yields the remainder when 13 is divided by 4, resulting in 1.

2.1.3   Congruence Classes

Congruence classes, also known as residue classes, are sets of integers that are congruent to each other modulo m. The congruence class of a modulo m is denoted as $[a]_m$, and it is the set of all integers b such that $a \equiv b \pmod{m}$.

For example, the congruence class of 2 modulo 4 is $[2]_4 = \{\ldots, -6, -2, 2, 6, \ldots\}$, as all these integers have a remainder of 2 when divided by 4.

2.1.4   Residue Systems

A residue system modulo m is a complete set of representatives for the congruence classes modulo m. It consists of exactly one representative from each congruence class. Common residue systems include the set of integers from 0 to $m - 1$, denoted as $\{0, 1, 2, \ldots, m - 1\}$.

2.1.5   Arithmetic in Modular Arithmetic

Arithmetic operations in modular arithmetic are performed on congruence classes. The sum, difference, and product of congruence classes are also congruence classes. However, division requires additional consideration, especially when finding the modular inverse.

Addition and Subtraction

The sum and difference of congruence classes are computed in the standard way, and the result is taken modulo m. For instance:

$$[5]_8 + [3]_8 = [8]_8 = [0]_8$$

Multiplication

The product of congruence classes is determined by multiplying the representatives and taking the result modulo m:

$$[4]_7 \times [3]_7 = [12]_7 = [5]_7$$

Division and Modular Inverse

Division in modular arithmetic involves finding the modular inverse. Given a congruence class $[a]_m$, the modular inverse x is an integer such that $ax \equiv 1 \pmod{m}$. The division is then computed as multiplication by the modular inverse:

$$[3]_{10} \nabla \cdot [7]_{10} = [3]_{10} \times [3]_{10}^{-1} = [9]_{10}$$

2.1.6 Applications of Modular Arithmetic

Modular arithmetic finds applications in diverse fields, including cryptography, computer science, and number theory. Its ability to handle cyclic patterns and remainders makes it a valuable tool in solving practical problems and designing efficient algorithms.

2.1.7 Cryptography

One of the prominent applications of modular arithmetic is in cryptography, where it forms the basis for various encryption and decryption algorithms. The security of public-key cryptosystems, such as RSA, relies on the difficulty of factoring large composite numbers, a problem well-suited for modular arithmetic techniques.

2.1.8   Computer Science

In computer science, modular arithmetic is widely used for addressing memory allocation and addressing issues. The cyclic nature of modular arithmetic is exploited in algorithms that involve circular buffers, hash functions, and addressing in data structures.

2.1.9   Number Theory

Modular arithmetic plays a central role in number theory, providing a powerful tool for exploring the properties of integers. It is extensively used in the study of prime numbers, Diophantine equations, and other number-theoretic problems.

2.1.10   Coding Theory

Coding theory, which deals with the design of error-detecting and error-correcting codes, relies heavily on modular arithmetic. Techniques such as checksums and cyclic redundancy checks involve modular arithmetic to ensure data integrity in communication systems.

2.1.11   Clock Arithmetic

The concept of clock arithmetic, where time is represented modulo 12 or 24, is a simple yet illustrative application of modular arithmetic. It demonstrates the cyclic nature of time representation, where 12 o'clock is congruent to 0 o'clock.

2.1.12   Chinese Remainder Theorem

The Chinese Remainder Theorem is a powerful result in modular arithmetic that provides a solution to a system of simultaneous congruences. It has applications in number theory, cryptography, and coding theory.

2.1.13 Fermat's Little Theorem

Fermat's Little Theorem is another fundamental result in modular arithmetic. It states that if p is a prime number and a is an integer not divisible by p, then $a^{p-1} \equiv 1 \pmod{p}$. This theorem finds applications in primality testing and modular exponentiation algorithms.

2.1.14 Euler's Totient Function

Euler's Totient Function is a crucial tool in modular arithmetic, providing the count of positive integers less than a given integer n that are coprime to n. It has applications in cryptography, particularly in the RSA algorithm.

2.1.15 Linear Diophantine Equations

Modular arithmetic is instrumental in solving linear Diophantine equations, which are equations of the form $ax + by = c$, where a, b, c, and the variables x and y are integers. Modular arithmetic techniques are applied to determine conditions for the existence of solutions and to find specific solutions.

2.1.16 Finite Fields

The concept of finite fields, also known as Galois fields, involves modular arithmetic and plays a significant role in coding theory and algebraic geometry. Finite fields are sets of numbers where arithmetic operations are performed modulo a prime number or a power of a prime.

2.1.17 Error Detection and Correction

Error-detecting and error-correcting codes, vital in data transmission and storage, rely heavily on modular arithmetic. Techniques such as Hamming codes and Reed-Solomon codes utilize the principles of modular arithmetic to detect and correct errors in transmitted data.

2.1.18   Coding Theory Applications

Modular arithmetic finds widespread applications in coding theory, a branch of information theory that deals with the encoding and decoding of information for reliable transmission. Linear block codes, cyclic codes, and convolutional codes are examples of coding techniques that leverage modular arithmetic.

2.1.19   Future Directions

As we progress through this book, our journey will extend beyond the basics of modular arithmetic into advanced topics and emerging research areas. The exploration of Chinese Remainder Theorem, Fermat's Little Theorem, and applications in cryptography and coding theory will pave the way for a deeper understanding of the subject.

2.1.20   Open Problems

The study of modular arithmetic is not without its challenges, and several open problems await exploration. From the existence of efficient algorithms for certain modular arithmetic computations to the deeper understanding of the interplay between modular arithmetic and other mathematical structures, these open problems present exciting avenues for future research.

2.1.21   Trends in Research

Current trends in research indicate a growing interest in the application of modular arithmetic in emerging technologies such as quantum computing, blockchain, and secure multiparty computation. Researchers are exploring novel algorithms and cryptographic protocols that leverage the unique properties of modular arithmetic for enhanced security and efficiency.

2.2   Modulo Operations

The core of modular arithmetic lies in the concept of modulo operations. Modulo, often denoted as $a \bmod m$ or $a \pmod m$, is the remainder when a is divided by m. This operation plays a pivotal role in defining congruence and exploring the cyclic nature of modular arithmetic.

2.2.1   Modulo Arithmetic

In its essence, modulo arithmetic is about working with remainders. The expression $a \equiv b \pmod m$ signifies that a and b have the same remainder when divided by m. This relation, called congruence, forms the foundation of modular arithmetic.

2.2.2   Basic Modulo Formulas

The basic modulo formulas are fundamental for understanding the behavior of modular arithmetic. For any integers a, b, and m, we have:

1. $a \equiv a \pmod m$

2. $a \equiv b \pmod m \implies b \equiv a \pmod m$

3. $a \equiv b \pmod m$ and $b \equiv c \pmod m \implies a \equiv c \pmod m$

4. $a + b \equiv (a \bmod m) + (b \bmod m) \pmod m$

5. $a - b \equiv (a \bmod m) - (b \bmod m) \pmod m$

6. $a \cdot b \equiv (a \bmod m) \cdot (b \bmod m) \pmod m$

These formulas encapsulate the principles of modular addition, subtraction, and multiplication.

2.2.3   Modulo and Remainder

The modulo operation is intrinsically connected to the remainder when dividing integers. For example, $15 \equiv 3 \pmod 4$ because 15 leaves a remainder of 3 when divided by 4.

2.2.4   Clock Arithmetic Revisited

To illustrate modulo arithmetic, consider the analogy of a clock. If it is currently 7 o'clock and we want to know the time 12 hours later, we can use modulo arithmetic:

$$(7 + 12) \bmod 12 = 7$$

So, 7 o'clock plus 12 hours is equivalent to 7 o'clock again, demonstrating the cyclic nature of modulo arithmetic.

2.2.5   Negative Numbers in Modulo

Dealing with negative numbers in modulo arithmetic requires careful consideration. The standard definition is $a \bmod m = a - m \left\lfloor \frac{a}{m} \right\rfloor$. For example, $-5 \bmod 3$ yields 1, as $-5 + 3 \times 2 = 1$.

2.2.6   Modulo Exponentiation

Modulo exponentiation is a crucial operation, especially in cryptography. The expression $a^b \pmod{m}$ represents the remainder when a raised to the power of b is divided by m. Efficient algorithms, like the square-and-multiply method, are employed for large numbers.

2.2.7   Modulo Inverses

The concept of modulo inverses plays a significant role in modular arithmetic. The modular inverse of a modulo m, denoted as a^{-1}, is the integer such that $a \cdot a^{-1} \equiv 1 \pmod{m}$. For example, $3^{-1} \equiv 7 \pmod{10}$ because $3 \cdot 7 \equiv 1 \pmod{10}$.

2.2.8 Modulo Operations in Coding

In computer programming, modulo operations are commonly used for various purposes. They are employed to determine if a number is even or odd (n mod 2), to wrap array indices within bounds, and to create repeating patterns in algorithms.

2.2.9 Modulo and Factorials

The connection between modulo and factorials is notable. The expression $n! \equiv 0$ (mod m) for $n \geq m$ due to the presence of m as a factor in the product.

2.2.10 Modulo and Prime Numbers

Modulo arithmetic is intimately linked with prime numbers. Fermat's Little Theorem states that for any prime p and a not divisible by p, $a^{p-1} \equiv 1$ (mod p). This theorem has significant applications in number theory and cryptography.

2.2.11 Cyclic Patterns in Modulo

One of the fascinating aspects of modulo arithmetic is its ability to create cyclic patterns. For instance, consider n^2 mod 4 for various values of n. The results exhibit a cyclic pattern: $0, 1, 0, 1, \ldots$, revealing the periodic nature of modulo operations.

2.2.12 Modulo and Quadratic Residues

Exploring quadratic residues in modulo arithmetic involves investigating values of a^2 (mod m) for different a and m. The study of quadratic residues is essential in number theory and cryptography.

2.2.13 Efficient Modulo Algorithms

Efficiency in modulo operations is crucial, particularly in applications involving large numbers. Algorithms like Barrett reduction and Montgomery reduction are employed to optimize modular arithmetic computations.

2.2.14   Modulo and Cryptographic Hash Functions

Cryptographic hash functions often utilize modulo operations to map input data
to a fixed-size output. The resulting hash value is typically computed as the
modulo of a large prime number.

2.2.15   Modulo and Group Theory

The principles of modulo arithmetic align with concepts in group theory. The
set of integers modulo m forms a group under addition, and when m is prime,
the set forms a finite field, a fundamental concept in algebraic structures.

2.2.16   Applications in Coding Theory

Modulo arithmetic finds applications in coding theory, particularly in the design
of error-detecting and error-correcting codes. Techniques such as cyclic redundancy checks and checksums leverage the properties of modulo operations.

2.3   Congruence

Congruence is a fundamental concept in modular arithmetic, forming the basis
for understanding the relationships between integers with respect to a modulus. This section explores the definition of congruence, its properties, and its
applications in various mathematical contexts.

2.3.1   Definition of Congruence

In modular arithmetic, two integers a and b are said to be congruent modulo
m, denoted as $a \equiv b \pmod{m}$, if they leave the same remainder when divided
by m. Mathematically, this can be expressed as:

$$a \equiv b \pmod{m} \iff m \text{ divides } (a - b)$$

For example, $17 \equiv 5 \pmod 6$ because $17 - 5 = 12$ is divisible by 6.

2.3.2 Properties of Congruence

Congruence exhibits several important properties that are crucial for its application in modular arithmetic:

- **Reflexivity:** $a \equiv a \pmod m$ for any integer a and modulus m.

- **Symmetry:** If $a \equiv b \pmod m$, then $b \equiv a \pmod m$.

- **Transitivity:** If $a \equiv b \pmod m$ and $b \equiv c \pmod m$, then $a \equiv c \pmod m$.

These properties make congruence an equivalence relation, providing a structured way to categorize integers into congruence classes.

2.3.3 Congruence Classes

Congruence classes, also known as residue classes, are sets of integers that are congruent to each other modulo m. The congruence class of a modulo m is denoted as $[a]_m$ and is defined as:

$$[a]_m = \{b \mid b \equiv a \pmod m\}$$

For instance, the congruence class of 2 modulo 4 is $[2]_4 = \{\ldots, -6, -2, 2, 6, \ldots\}$, as all these integers have a remainder of 2 when divided by 4.

2.3.4 Modular Arithmetic and Clock Arithmetic

Congruence finds an intuitive interpretation in clock arithmetic. If we think of numbers on a clock, two numbers are congruent modulo 12 if they point to the same position on the clock. For example, 7 and 19 are congruent modulo 12 since they both represent the same hour on the clock.

2.3.5   Arithmetic Properties of Congruence

Congruence preserves the results of basic arithmetic operations, forming the basis for modular arithmetic:

- **Addition:** If $a \equiv b \pmod{m}$ and $c \equiv d \pmod{m}$, then $a + c \equiv b + d \pmod{m}$.

- **Subtraction:** If $a \equiv b \pmod{m}$ and $c \equiv d \pmod{m}$, then $a - c \equiv b - d \pmod{m}$.

- **Multiplication:** If $a \equiv b \pmod{m}$ and $c \equiv d \pmod{m}$, then $a \cdot c \equiv b \cdot d \pmod{m}$.

These properties allow us to perform operations on congruence classes rather than individual numbers.

2.3.6   Applications of Congruence

Congruence is a versatile concept with applications in various mathematical domains. Some notable applications include:

Number Theory

In number theory, congruence is extensively used to study properties of integers. It plays a central role in the exploration of prime numbers, divisibility, and Diophantine equations.

Cryptography

Congruence forms the foundation of many cryptographic algorithms. Public-key cryptosystems like RSA rely on the difficulty of factoring large composite numbers, a problem that involves congruence.

Computer Science

In computer science, congruence is employed in addressing memory allocation and indexing in data structures. It ensures efficient use of finite resources and facilitates cyclic patterns in algorithms.

Coding Theory

Coding theory, which deals with the design of error-detecting and error-correcting codes, leverages congruence to create efficient encoding and decoding mechanisms. Cyclic codes, in particular, rely on congruence properties.

2.3.7 Congruence and Modular Inverses

The concept of a modular inverse is closely tied to congruence. If $a \equiv b \pmod{m}$ and b has a modular inverse, then a also has the same modular inverse. The modular inverse of b is denoted as b^{-1}, and it satisfies $b \cdot b^{-1} \equiv 1 \pmod{m}$.

2.3.8 Chinese Remainder Theorem

The Chinese Remainder Theorem is a powerful application of congruence. It provides a solution to a system of simultaneous congruences, allowing us to find a unique solution modulo the product of the moduli involved.

2.3.9 Fermat's Little Theorem

Fermat's Little Theorem is a significant result in number theory that relies on congruence. It states that if p is a prime number and a is an integer not divisible by p, then $a^{p-1} \equiv 1 \pmod{p}$.

2.3.10 Linear Congruences

Linear congruences, of the form $ax \equiv b \pmod{m}$, are commonly encountered in various mathematical problems. Solving linear congruences involves finding values of x that satisfy the congruence relation.

2.3.11 Congruence and Quadratic Residues

The study of quadratic residues involves exploring congruences of the form $x^2 \equiv a \pmod{m}$. Understanding these congruences is crucial in number theory, cryptography, and algebraic structures.

2.3.12 Residue Number System

The residue number system (RNS) is a specialized application of congruence. It represents integers using their residues with respect to a set of pairwise coprime moduli, allowing for efficient arithmetic operations.

Chapter 3

Arithmetic Operations

3.1 Addition and Subtraction

The fundamental operations of addition and subtraction play a crucial role in modular arithmetic, providing the basis for exploring the cyclic nature of integers under modular conditions. This section delves into the rules, properties, and applications of addition and subtraction in the context of modular arithmetic.

3.1.1 Modular Addition

In modular addition, the sum of two integers a and b is calculated modulo m, denoted as $a + b \pmod{m}$. The formula for modular addition is:

$$a + b \equiv (a \pmod{m}) + (b \pmod{m}) \pmod{m}$$

This formula ensures that the result remains within the range of 0 to $m - 1$, creating a cyclic pattern.

3.1.2   Modular Subtraction

Similarly, modular subtraction involves finding the difference of two integers a and b modulo m, denoted as $a - b \pmod{m}$. The formula for modular subtraction is:

$$a - b \equiv (a \pmod{m}) - (b \pmod{m}) \pmod{m}$$

This formula accounts for the cyclical nature of modular arithmetic, ensuring the result remains within the specified modulus.

3.1.3   Addition Examples

Let's consider some examples to illustrate modular addition. Suppose we have $a = 17$, $b = 8$, and $m = 5$. The modular sum $17 + 8 \pmod 5$ is calculated as:

$$17 + 8 \equiv (17 \pmod 5) + (8 \pmod 5) \pmod 5 \equiv 2 + 3 \pmod 5 \equiv 0 \pmod 5$$

Therefore, $17 + 8 \equiv 0 \pmod 5$. This result showcases how modular addition respects the cyclical nature of arithmetic modulo 5.

3.1.4   Subtraction Examples

Now, let's explore modular subtraction with $a = 10$, $b = 4$, and $m = 7$. The modular difference $10 - 4 \pmod 7$ is calculated as:

$$10 - 4 \equiv (10 \pmod 7) - (4 \pmod 7) \pmod 7 \equiv 3 - 4 \pmod 7 \equiv -1 \pmod 7$$

$$\equiv 6 \pmod 7$$

Hence, $10 - 4 \equiv 6 \pmod 7$. This example demonstrates how modular subtraction handles negative results within the specified modulus.

3.1.5 Properties of Modular Addition and Subtraction

Modular addition and subtraction exhibit properties that distinguish them in the realm of modular arithmetic:

- **Commutativity:** $a + b \equiv b + a \pmod{m}$

- **Associativity:** $a + (b + c) \equiv (a + b) + c \pmod{m}$

- **Additive Identity:** $a + 0 \equiv a \pmod{m}$

- **Additive Inverse:** $a + (-a) \equiv 0 \pmod{m}$

These properties align with the familiar rules of arithmetic and contribute to the coherence of modular addition and subtraction.

3.1.6 Applications in Clock Arithmetic

The principles of modular addition and subtraction find a natural analogy in clock arithmetic. Consider a clock with 12 hours, where $11 + 2$ and $1 - 8$ are equivalent to 1 and 5 o'clock, respectively. The cyclic nature of clock arithmetic mirrors the behavior of modular addition and subtraction.

3.1.7 Efficient Computation

One of the advantages of modular arithmetic is its efficiency in computation. Modular addition and subtraction can be implemented using basic arithmetic operations, making them computationally straightforward.

3.1.8 Modular Addition and Subtraction in Coding

In computer science, modular addition and subtraction are extensively used for various purposes. They are employed in algorithms that involve circular buffers, addressing in data structures, and creating repeating patterns.

3.1.9   Extended Modular Addition

Extended modular addition is a concept that extends the range of modular results beyond 0 to $m - 1$. If $a + b$ exceeds $m - 1$, we can wrap around to the beginning of the cycle. For example, in (mod 7), $5 + 4$ results in 2 rather than 9.

3.1.10   Extended Modular Subtraction

Similarly, extended modular subtraction accommodates negative results by wrapping around to the end of the cycle. In (mod 6), $2 - 5$ results in 3 instead of -3.

3.1.11   Modular Arithmetic Clock Diagram

A helpful visualization for modular addition and subtraction is a modular arithmetic clock diagram. This diagram illustrates the cyclical nature of modular operations, providing a geometric representation of arithmetic modulo m.

3.1.12   Congruence and Modular Addition

The connection between congruence and modular addition is significant. If $a \equiv b \pmod{m}$, then $a + c \equiv b + c \pmod{m}$. This property allows us to perform operations on congruence classes.

3.1.13   Congruence and Modular Subtraction

Similarly, congruence and modular subtraction are intertwined. If $a \equiv b \pmod{m}$, then $a - c \equiv b - c \pmod{m}$. The principles of congruence extend naturally to modular arithmetic.

3.1.14 Modular Addition and Subtraction in Cryptography

In cryptography, modular addition and subtraction are foundational operations. They play a crucial role in encryption and decryption algorithms, ensuring the security and integrity of cryptographic systems.

3.2 Multiplication

Multiplication is a fundamental operation in modular arithmetic, allowing us to explore the cyclic patterns and structures inherent in the integers under modular conditions. In this section, we delve into the rules, properties, and applications of multiplication in the context of modular arithmetic.

3.2.1 Modular Multiplication

In modular multiplication, the product of two integers a and b is calculated modulo m, denoted as $a \cdot b \pmod{m}$. The formula for modular multiplication is:

$$a \cdot b \equiv (a \pmod{m}) \cdot (b \pmod{m}) \pmod{m}$$

This formula ensures that the result remains within the range of 0 to $m - 1$, preserving the cyclical nature of modular arithmetic.

3.2.2 Multiplicative Identity

Similar to modular addition and subtraction, modular multiplication has a multiplicative identity. For any integer a, $a \cdot 1 \equiv a \pmod{m}$.

3.2.3 Multiplicative Inverse

The concept of a multiplicative inverse is significant in modular multiplication. If a has a modular inverse a^{-1} modulo m, then $a \cdot a^{-1} \equiv 1 \pmod{m}$.

3.2.4   Multiplication Examples

Let's consider some examples to illustrate modular multiplication. Suppose we have $a = 3$, $b = 4$, and $m = 7$. The modular product $3 \cdot 4 \pmod 7$ is calculated as:

$$3 \cdot 4 \equiv (3 \pmod 7) \cdot (4 \pmod 7) \pmod 7 \equiv 3 \cdot 4 \pmod 7 \equiv 5 \pmod 7$$

Therefore, $3 \cdot 4 \equiv 5 \pmod 7$. This example demonstrates how modular multiplication respects the cyclical nature of arithmetic modulo 7.

3.2.5   Properties of Modular Multiplication

Modular multiplication exhibits several properties that distinguish it within the realm of modular arithmetic:

- **Commutativity:** $a \cdot b \equiv b \cdot a \pmod m$

- **Associativity:** $a \cdot (b \cdot c) \equiv (a \cdot b) \cdot c \pmod m$

- **Multiplicative Identity:** $a \cdot 1 \equiv a \pmod m$

- **Multiplicative Inverse:** If a has a multiplicative inverse a^{-1}, then $a \cdot a^{-1} \equiv 1 \pmod m$

- **Distributive Property:** $a \cdot (b + c) \equiv a \cdot b + a \cdot c \pmod m$

These properties align with the familiar rules of arithmetic and contribute to the coherence of modular multiplication.

3.2.6   Applications in Cryptography

Modular multiplication plays a crucial role in cryptographic algorithms. Public-key cryptosystems often involve large modular multiplication operations, and the difficulty of factoring such products contributes to the security of these systems.

3.2.7 Efficient Computation

Similar to modular addition and subtraction, modular multiplication can be efficiently computed using basic arithmetic operations. This efficiency is particularly valuable in applications involving large numbers.

3.2.8 Modular Exponentiation

Modular exponentiation extends the concept of modular multiplication to the realm of raising a number to a power modulo m. The expression $a^b \pmod{m}$ represents the remainder when a raised to the power of b is divided by m.

3.2.9 Applications in Coding Theory

In coding theory, modular multiplication is utilized in the design of error-detecting and error-correcting codes. Techniques such as Reed-Solomon codes leverage the properties of modular multiplication to achieve robust data transmission.

3.2.10 Chinese Remainder Theorem Revisited

The Chinese Remainder Theorem, introduced earlier, involves modular multiplication as part of its solution to systems of simultaneous congruences. It provides a method to find a unique solution modulo the product of the moduli involved.

3.2.11 Linear Congruences and Modular Multiplication

Linear congruences, of the form $ax \equiv b \pmod{m}$, often involve modular multiplication. Solving such congruences requires considering the modular inverse of a, which, in turn, involves modular multiplication.

3.2.12   Efficient Modular Multiplication Algorithms

Efficiency in modular multiplication is crucial, especially when dealing with large numbers. Algorithms like Montgomery multiplication and Karatsuba multiplication are employed to optimize modular multiplication computations.

3.2.13   Modular Multiplication and Quadratic Residues

The study of quadratic residues involves exploring congruences of the form $x^2 \equiv a \pmod{m}$. Modular multiplication plays a crucial role in investigating and understanding quadratic residues.

3.2.14   Applications in Number Theory

In number theory, modular multiplication is central to exploring properties of integers and their relationships under modular conditions. It is particularly useful in investigating divisibility and prime numbers.

3.2.15   Multiplication in the Residue Number System

The Residue Number System (RNS) is an application of modular multiplication. It represents integers using their residues with respect to a set of pairwise coprime moduli, allowing for efficient arithmetic operations.

3.3   Division

Division is a crucial operation in modular arithmetic, offering a way to find the quotient and remainder when one integer is divided by another modulo m. In this section, we explore the rules, properties, and applications of division in the context of modular arithmetic.

3.3.1 Modular Division

In modular division, we aim to find the result of dividing one integer a by another b modulo m, denoted as $a \nabla \cdot b \pmod{m}$. The formula for modular division is not as straightforward as addition, subtraction, and multiplication, and it relies on the concept of the modular inverse.

$$a \nabla \cdot b \equiv a \cdot b^{-1} \pmod{m}$$

Here, b^{-1} is the modular inverse of b modulo m, satisfying $b \cdot b^{-1} \equiv 1 \pmod{m}$.

3.3.2 Modular Inverse

The modular inverse b^{-1} exists if b is relatively prime to m, meaning that b and m share no common factors except 1. If b and m are not relatively prime, the modular inverse does not exist.

3.3.3 Division Examples

Let's consider an example to illustrate modular division. Suppose we have $a = 10$, $b = 3$, and $m = 7$. We want to find $10 \nabla \cdot 3 \pmod{7}$. First, we need to find the modular inverse of 3 modulo 7, denoted as 3^{-1}. In this case, $3^{-1} \equiv 5 \pmod{7}$ because $3 \cdot 5 \equiv 1 \pmod{7}$.

Now, we can calculate the modular division:

$$10 \nabla \cdot 3 \equiv 10 \cdot 3^{-1} \pmod{7} \equiv 10 \cdot 5 \pmod{7} \equiv 50 \pmod{7} \equiv 1 \pmod{7}$$

Therefore, $10 \nabla \cdot 3 \equiv 1 \pmod{7}$.

3.3.4 Properties of Modular Division

Modular division has some unique properties that stem from the properties of the modular inverse:

- **Existence of Inverse:** b^{-1} exists if b is relatively prime to m.

- **Multiplicative Identity:** $a\nabla \cdot 1 \equiv a \pmod{m}$

- **Cancellation:** If $a\nabla \cdot b \equiv a\nabla \cdot c \pmod{m}$, then $b \equiv c \pmod{m}$

These properties highlight the nuanced nature of modular division and its reliance on the modular inverse.

3.3.5   Applications in Cryptography

Modular division plays a crucial role in cryptographic algorithms, especially in public-key cryptosystems. The computation of modular inverses is essential for tasks such as key generation and decryption.

3.3.6   Efficient Computation of Modular Inverse

The computation of the modular inverse can be performed using various algorithms, such as the Extended Euclidean Algorithm. Efficient algorithms are crucial in cryptographic applications involving large numbers.

3.3.7   Chinese Remainder Theorem and Modular Division

The Chinese Remainder Theorem, introduced earlier, involves solving systems of simultaneous congruences. Modular division, particularly finding the modular inverse, is a key step in this process.

3.3.8   Division in the Residue Number System

In the Residue Number System (RNS), modular division is a fundamental operation. Representing numbers using their residues with respect to a set of pairwise coprime moduli involves both modular multiplication and division.

3.3.9 Division and Quadratic Residues

The study of quadratic residues involves exploring congruences of the form $x^2 \equiv a \pmod{m}$. Modular division is often employed in solving equations related to quadratic residues.

3.3.10 Linear Congruences and Modular Division

Linear congruences, of the form $ax \equiv b \pmod{m}$, may involve modular division when trying to isolate x. Solving such equations requires finding the modular inverse of a.

3.3.11 Applications in Coding Theory

In coding theory, modular division is utilized in error-detecting and error-correcting codes. Techniques such as Reed-Solomon codes leverage the properties of modular division to achieve robust data transmission.

3.3.12 Division and Fermat's Little Theorem

Fermat's Little Theorem, a significant result in number theory, involves modular division. It states that if p is a prime number and a is an integer not divisible by p, then $a^{p-1} \equiv 1 \pmod{p}$. The proof of this theorem often relies on modular division.

3.3.13 Division in Clock Arithmetic

In clock arithmetic, modular division allows us to find, for example, the hour on a clock given the total number of hours. If h represents the total hours and d is the divisor (number of hours in a day), then $h \nabla \cdot d$ gives us the hour.

3.3.14 Extended Modular Division

Extended modular division, similar to extended modular addition and subtraction, is a concept that extends the range of modular results beyond 0 to $m - 1$.

It accommodates cases where the division result is negative.

Chapter 4

Properties of Modular Arithmetic

4.1 Associativity

Associativity is a fundamental property of modular arithmetic that governs how operations, such as addition and multiplication, are performed on integers modulo m. In this section, we delve into the concept of associativity, its significance, and its applications in the context of modular arithmetic.

4.1.1 Associativity in Modular Addition

Associativity in modular addition dictates that the grouping of numbers does not affect the result when adding them modulo m. Mathematically, this is expressed as:

$$(a + b) + c \equiv a + (b + c) \pmod{m}$$

This property ensures that the outcome of modular addition is consistent regardless of the grouping of terms.

4.1.2 Associativity Examples in Modular Addition

Let's consider an example to illustrate associativity in modular addition. Suppose we have $a = 5$, $b = 7$, and $c = 3$, with a modulus $m = 10$. We want to evaluate $(5 + 7) + 3$ and compare it with $5 + (7 + 3)$ modulo 10.

$$(5 + 7) + 3 \equiv 12 \pmod{10} \equiv 2 \pmod{10}$$

$$5 + (7 + 3) \equiv 5 + 10 \pmod{10} \equiv 5 \pmod{10}$$

As expected, both expressions yield the same result, confirming the associativity of modular addition.

4.1.3 Associativity in Modular Multiplication

Associativity extends to modular multiplication as well. For any integers a, b, and c modulo m, the following holds:

$$(a \cdot b) \cdot c \equiv a \cdot (b \cdot c) \pmod{m}$$

This property ensures that the result of modular multiplication remains consistent regardless of the grouping of factors.

4.1.4 Associativity Examples in Modular Multiplication

Consider an example to illustrate associativity in modular multiplication. Let's take $a = 4$, $b = 6$, and $c = 2$ with a modulus $m = 7$. We want to evaluate $(4 \cdot 6) \cdot 2$ and compare it with $4 \cdot (6 \cdot 2)$ modulo 7.

$$(4 \cdot 6) \cdot 2 \equiv 24 \pmod{7} \equiv 3 \pmod{7}$$

$$4 \cdot (6 \cdot 2) \equiv 4 \cdot 12 \pmod{7} \equiv 4 \pmod{7}$$

Again, we observe that both expressions yield the same result, confirming the associativity of modular multiplication.

4.1.5 Associativity of Exponentiation

Associativity further extends to exponentiation in modular arithmetic. For any integers a, b, and c modulo m, the following holds:

$$(a^b)^c \equiv a^{(b \cdot c)} \pmod{m}$$

This property allows for consistent results when raising a number to a power modulo m regardless of the grouping of exponents.

4.1.6 Associativity Examples in Exponentiation

Consider an example to illustrate associativity in exponentiation. Let $a = 2$, $b = 3$, and $c = 4$ with a modulus $m = 5$. We want to evaluate $(2^3)^4$ and compare it with $2^{(3 \cdot 4)}$ modulo 5.

$$(2^3)^4 \equiv 8^4 \pmod{5} \equiv 2^4 \pmod{5} \equiv 16 \pmod{5} \equiv 1 \pmod{5}$$

$$2^{(3 \cdot 4)} \equiv 2^{12} \pmod{5} \equiv 16 \pmod{5} \equiv 1 \pmod{5}$$

As expected, both expressions yield the same result, affirming the associativity of exponentiation in modular arithmetic.

4.1.7 Associativity in Residue Number System

The Residue Number System (RNS) leverages the associativity of modular arithmetic. Operations in the RNS involve modular addition and multiplication, and the associativity property ensures consistent results in the encoding and decoding processes.

4.1.8 Associativity and Cryptography

Associativity is a desirable property in cryptographic algorithms. It ensures that the results of operations remain consistent, contributing to the reliability and security of cryptographic systems.

4.1.9　Associativity in Clock Arithmetic

Associativity in modular arithmetic finds a practical application in clock arithmetic. Consider a clock with 12 hours. The associativity property ensures that adding hours in different orders yields the same final time.

4.1.10　Associativity in Computer Science

In computer science, associativity is a crucial consideration in algorithms and data structures. Modular arithmetic, with its associativity property, is employed in various applications such as hashing functions and error-checking algorithms.

4.1.11　Associativity and Linear Congruences

Linear congruences, of the form $ax \equiv b \pmod{m}$, often involve modular addition and multiplication. The associativity property ensures that the solutions to linear congruences are well-defined and consistent.

4.1.12　Associativity and Polynomial Arithmetic

Associativity plays a role in polynomial arithmetic over finite fields, where coefficients are taken modulo a prime number. The associativity property ensures consistent results when adding and multiplying polynomials.

4.1.13　Associativity and Group Theory

Associativity is a key property in group theory, and modular arithmetic exhibits a group structure under addition and multiplication modulo m. The associativity property is fundamental to the group axioms.

4.1.14 Associativity and Coding Theory

In coding theory, associativity is crucial in designing error-detecting and error-correcting codes. Operations involving modular addition and multiplication rely on the associativity property to ensure the integrity of transmitted data.

4.1.15 Associativity in Quadratic Residues

The study of quadratic residues involves modular multiplication, and the associativity property ensures consistent results when exploring quadratic residues modulo m.

4.2 Commutativity

Commutativity is a fundamental property of modular arithmetic that governs the order of operations, such as addition and multiplication, on integers modulo m. In this section, we explore the concept of commutativity, its mathematical expressions, and its implications within the context of modular arithmetic.

4.2.1 Commutativity in Modular Addition

Commutativity in modular addition asserts that changing the order of the terms does not affect the result when adding them modulo m. Mathematically, this is expressed as:

$$a + b \equiv b + a \pmod{m}$$

This property ensures that the outcome of modular addition is consistent regardless of the order in which the numbers are added.

4.2.2 Commutativity Examples in Modular Addition

Let's consider an example to illustrate commutativity in modular addition. Suppose we have $a = 8$, $b = 5$, and a modulus $m = 7$. We want to evaluate $8 + 5$

and compare it with $5 + 8$ modulo 7.

$$8 + 5 \equiv 13 \pmod{7} \equiv 6 \pmod{7}$$

$$5 + 8 \equiv 13 \pmod{7} \equiv 6 \pmod{7}$$

As expected, both expressions yield the same result, confirming the commutativity of modular addition.

4.2.3   Commutativity in Modular Multiplication

Commutativity extends to modular multiplication as well. For any integers a and b modulo m, the following holds:

$$a \cdot b \equiv b \cdot a \pmod{m}$$

This property ensures that changing the order of factors does not affect the result when multiplying them modulo m.

4.2.4   Commutativity Examples in Modular Multiplication

Consider an example to illustrate commutativity in modular multiplication. Let $a = 3$, $b = 6$, and a modulus $m = 7$. We want to evaluate $3 \cdot 6$ and compare it with $6 \cdot 3$ modulo 7.

$$3 \cdot 6 \equiv 18 \pmod{7} \equiv 4 \pmod{7}$$

$$6 \cdot 3 \equiv 18 \pmod{7} \equiv 4 \pmod{7}$$

Once again, both expressions yield the same result, confirming the commutativity of modular multiplication.

4.2.5 Commutativity of Exponentiation

Commutativity further extends to exponentiation in modular arithmetic. For any integers a and b modulo m, the following holds:

$$a^b \equiv b^a \pmod{m}$$

This property allows for consistent results when raising a number to a power modulo m, regardless of the order of the base and exponent.

4.2.6 Commutativity Examples in Exponentiation

Consider an example to illustrate commutativity in exponentiation. Let $a = 2$, $b = 4$, and a modulus $m = 5$. We want to evaluate 2^4 and compare it with 4^2 modulo 5.

$$2^4 \equiv 16 \pmod{5} \equiv 1 \pmod{5}$$

$$4^2 \equiv 16 \pmod{5} \equiv 1 \pmod{5}$$

As expected, both expressions yield the same result, confirming the commutativity of exponentiation in modular arithmetic.

4.2.7 Commutativity in Residue Number System

The Residue Number System (RNS) leverages the commutativity of modular arithmetic. Operations in the RNS involve modular addition and multiplication, and the commutativity property ensures consistent results in the encoding and decoding processes.

4.2.8 Commutativity and Cryptography

Commutativity is a desirable property in cryptographic algorithms. It ensures that the results of operations remain consistent, contributing to the reliability and security of cryptographic systems.

4.2.9   Commutativity in Clock Arithmetic

Commutativity in modular arithmetic finds a practical application in clock arithmetic. Consider a clock with 12 hours. The commutativity property ensures that the order in which hours are added does not affect the final time.

4.2.10   Commutativity in Computer Science

In computer science, commutativity is a crucial consideration in algorithms and data structures. Modular arithmetic, with its commutativity property, is employed in various applications such as hashing functions and error-checking algorithms.

4.2.11   Commutativity and Linear Congruences

Linear congruences, of the form $ax \equiv b \pmod{m}$, often involve modular addition and multiplication. The commutativity property ensures that the solutions to linear congruences are well-defined and consistent.

4.2.12   Commutativity and Polynomial Arithmetic

Commutativity plays a role in polynomial arithmetic over finite fields, where coefficients are taken modulo a prime number. The commutativity property ensures consistent results when adding and multiplying polynomials.

4.2.13   Commutativity and Group Theory

Commutativity is a key property in group theory, and modular arithmetic exhibits a group structure under addition and multiplication modulo m. The commutativity property is fundamental to the group axioms.

4.2.14   Commutativity and Coding Theory

In coding theory, commutativity is crucial in designing error-detecting and error-correcting codes. Operations involving modular addition and multiplication rely on the commutativity property to ensure the integrity of transmitted data.

4.2.15 Commutativity in Quadratic Residues

The study of quadratic residues involves modular multiplication, and the commutativity property ensures consistent results when exploring quadratic residues modulo m.

4.3 Distributive Property

The distributive property is a fundamental aspect of modular arithmetic that governs the combination of addition and multiplication operations. In this section, we explore the distributive property, examine its mathematical expressions, and discuss its implications within the context of modular arithmetic.

4.3.1 Distributive Property in Modular Addition and Multiplication

The distributive property in modular arithmetic expresses the relationship between addition and multiplication. For any integers a, b, and c modulo m, the property can be stated as:

$$a \cdot (b + c) \equiv (a \cdot b) + (a \cdot c) \pmod{m}$$

This property allows for the distribution of the factor a across the sum of b and c in a modular arithmetic setting.

4.3.2 Distributive Property Example

Let's consider an example to illustrate the distributive property. Suppose we have $a = 3$, $b = 4$, and $c = 5$ with a modulus $m = 7$. We want to evaluate $3 \cdot (4 + 5)$ and compare it with $(3 \cdot 4) + (3 \cdot 5)$ modulo 7.

$$3 \cdot (4 + 5) \equiv 3 \cdot 9 \pmod{7} \equiv 27 \pmod{7} \equiv 6 \pmod{7}$$

$$(3 \cdot 4) + (3 \cdot 5) \equiv 12 + 15 \pmod 7 \equiv 27 \pmod 7 \equiv 6 \pmod 7$$

As expected, both expressions yield the same result, confirming the distributive property in modular arithmetic.

4.3.3   Extended Distributive Property

The distributive property can be extended to more than two terms. For any integers a, b, c, and d modulo m, the extended distributive property can be expressed as:

$$a \cdot (b + c + d) \equiv (a \cdot b) + (a \cdot c) + (a \cdot d) \pmod m$$

This extension allows for the distribution of the factor a across the sum of multiple terms in a modular arithmetic setting.

4.3.4   Extended Distributive Property Example

Consider an example to illustrate the extended distributive property. Let $a = 2$, $b = 3$, $c = 4$, and $d = 5$ with a modulus $m = 6$. We want to evaluate $2 \cdot (3+4+5)$ and compare it with $(2 \cdot 3) + (2 \cdot 4) + (2 \cdot 5)$ modulo 6.

$$2 \cdot (3 + 4 + 5) \equiv 2 \cdot 12 \pmod 6 \equiv 24 \pmod 6 \equiv 0 \pmod 6$$

$$(2 \cdot 3) + (2 \cdot 4) + (2 \cdot 5) \equiv 6 + 8 + 10 \pmod 6 \equiv 24 \pmod 6 \equiv 0 \pmod 6$$

Once again, both expressions yield the same result, confirming the extended distributive property in modular arithmetic.

4.3.5   Applications of the Distributive Property

The distributive property plays a crucial role in various mathematical and computational applications. Some notable applications include:

Polynomial Arithmetic

In polynomial arithmetic over finite fields, the distributive property is utilized when multiplying polynomials by distributing each term.

Matrix Operations

In matrix multiplication, the distributive property is essential. The product of a matrix and a sum of matrices is distributed across each term of the sum.

Coding Theory

In coding theory, the distributive property is employed when designing error-detecting and error-correcting codes. Operations involving encoding and decoding rely on the distributive property.

Linear Congruences

The distributive property is relevant in solving linear congruences, where it aids in simplifying expressions involving modular arithmetic.

Computer Algorithms

In computer algorithms, the distributive property is fundamental in optimizing computations. The property is leveraged in various algorithms, including those used in cryptography and data compression.

Factorization Techniques

The distributive property is often used in factorization techniques, where expressions are factored by distributing common factors.

Number Theory

In number theory, the distributive property is applied to simplify expressions involving modular arithmetic, facilitating the study of congruences.

4.3.6   Distributive Property and Modular Exponentiation

The distributive property extends to modular exponentiation. For any integers a, b, and c modulo m, the property can be expressed as:

$$a^{(b+c)} \equiv (a^b) \cdot (a^c) \pmod{m}$$

This property allows for the distribution of the base a across the sum of exponents b and c in a modular arithmetic setting.

4.3.7   Distributive Property Example in Modular Exponentiation

Consider an example to illustrate the distributive property in modular exponentiation. Let $a = 2$, $b = 3$, and $c = 4$ with a modulus $m = 5$. We want to evaluate $2^{(3+4)}$ and compare it with $2^3 \cdot 2^4$ modulo 5.

$$2^{(3+4)} \equiv 2^7 \pmod{5} \equiv 128 \pmod{5} \equiv 3 \pmod{5}$$

$$(2^3) \cdot (2^4) \equiv 8 \cdot 16 \pmod{5} \equiv 128 \pmod{5} \equiv 3 \pmod{5}$$

As anticipated, both expressions yield the same result, affirming the distributive property in modular exponentiation.

4.4   Cancellation Property

The cancellation property is a fundamental characteristic of modular arithmetic that allows for the simplification of equations by canceling common factors. In this section, we explore the cancellation property, examine its mathematical expressions, and discuss its implications within the context of modular arithmetic.

4.4.1 Cancellation Property in Modular Addition

The cancellation property in modular addition states that if $a + b \equiv a + c$ (mod m), then $b \equiv c$ (mod m). In other words, if two sums are congruent modulo m with a common term, the remaining terms are congruent modulo m.

4.4.2 Cancellation Property Example in Modular Addition

Let's consider an example to illustrate the cancellation property in modular addition. Suppose we have $a = 5$, $b = 3$, $c = 8$, and a modulus $m = 6$. If $5 + 3 \equiv 5 + 8$ (mod 6), then by the cancellation property, we can deduce that $3 \equiv 8$ (mod 6).

4.4.3 Cancellation Property in Modular Multiplication

The cancellation property also applies to modular multiplication. If $a \cdot b \equiv a \cdot c$ (mod m) and a is relatively prime to m, then $b \equiv c$ (mod m). This means that if two products are congruent modulo m with a common factor a that is relatively prime to m, the remaining factors are congruent modulo m.

4.4.4 Cancellation Property Example in Modular Multiplication

Consider an example to illustrate the cancellation property in modular multiplication. Let $a = 3$, $b = 4$, $c = 5$, and a modulus $m = 7$. If $3 \cdot 4 \equiv 3 \cdot 5$ (mod 7), then by the cancellation property, we can deduce that $4 \equiv 5$ (mod 7).

4.4.5 Cancellation Property and Modular Division

The cancellation property extends to modular division when a modular inverse exists. If $a \cdot x \equiv b \cdot x$ (mod m), and x is the modular inverse of a modulo m, then $a \equiv b$ (mod m). This implies that if the product of a number and a modular inverse is congruent modulo m, the numbers themselves are congruent modulo m.

4.4.6   Cancellation Property Example in Modular Division

Consider an example to illustrate the cancellation property in modular division. Let $a = 4$, $b = 8$, and $m = 7$. If $4 \cdot x \equiv 8 \cdot x \pmod 7$, where x is the modular inverse of 4 modulo 7, then by the cancellation property, we can deduce that $4 \equiv 8 \pmod 7$.

4.4.7   Cancellation Property in Linear Congruences

The cancellation property is particularly useful in solving linear congruences. If $ax \equiv bx \pmod m$, where a is relatively prime to m, then x can be canceled, and we obtain $a \equiv b \pmod m$.

4.4.8   Cancellation Property Example in Linear Congruences

Consider an example to illustrate the cancellation property in linear congruences. Let $a = 5$, $b = 2$, and $m = 6$. If $5x \equiv 2x \pmod 6$, where x is relatively prime to 6, then by the cancellation property, we can deduce that $5 \equiv 2 \pmod 6$.

4.4.9   Cancellation Property and Polynomial Congruences

The cancellation property extends to polynomial congruences. If $P(x) \equiv Q(x)$ $\pmod m$ and a is relatively prime to m, then $P(a) \equiv Q(a) \pmod m$. This implies that if two polynomials are congruent modulo m, their values at a given point a are also congruent modulo m.

4.4.10 Cancellation Property Example in Polynomial Congruences

Consider an example to illustrate the cancellation property in polynomial congruences. Let $P(x) = x^2 + 3x + 2$, $Q(x) = x^2 + 5x + 6$, $a = 4$, and $m = 7$. If $P(x) \equiv Q(x) \pmod{7}$, then by the cancellation property, we can deduce that $P(4) \equiv Q(4) \pmod{7}$.

4.4.11 Cancellation Property and Quadratic Residues

The cancellation property is relevant in the study of quadratic residues. If $a \equiv b \pmod{p}$, where p is an odd prime, then $a^2 \equiv b^2 \pmod{p}$. This property allows for the cancellation of terms in quadratic residue equations.

4.4.12 Cancellation Property Example in Quadratic Residues

Consider an example to illustrate the cancellation property in quadratic residues. Let $a = 3$, $b = 10$, and $p = 7$. If $3 \equiv 10 \pmod{7}$, then by the cancellation property, we can deduce that $3^2 \equiv 10^2 \pmod{7}$.

4.4.13 Cancellation Property in Group Theory

In group theory, the cancellation property is a defining characteristic of groups. Modular arithmetic under addition modulo m forms a group, and the cancellation property ensures that elements within the group can be canceled.

4.4.14 Cancellation Property and Unique Factorization

The cancellation property is closely related to the unique factorization property in modular arithmetic. The ability to cancel common factors is intricately connected to the prime factorization of integers modulo m.

4.4.15   Cancellation Property and Cryptography

The cancellation property plays a role in cryptographic protocols that involve modular arithmetic. It contributes to the efficiency and security of cryptographic algorithms.

Chapter 5

Applications

5.1 Cryptography

Cryptography, the art and science of secure communication, heavily relies on modular arithmetic to ensure the confidentiality, integrity, and authenticity of information. In this section, we delve into the applications of modular arithmetic in the field of cryptography, exploring fundamental concepts, mathematical formulas, and practical examples.

5.1.1 Basic Concepts in Cryptography

Cryptography involves various key concepts, including encryption, decryption, key generation, and secure communication protocols. At its core, the field aims to transform information into a format that is unreadable to unauthorized individuals, ensuring that only authorized parties can access the original data.

5.1.2 Mathematical Foundation of Cryptography

Modular arithmetic provides a robust mathematical foundation for many cryptographic algorithms. The use of modular arithmetic operations, such as modular addition, multiplication, and exponentiation, forms the basis of encryption and decryption processes.

5.1.3   Modular Exponentiation in Cryptography

One of the essential operations in cryptography is modular exponentiation. For a given base a, exponent b, and modulus m, the modular exponentiation is calculated as:

$$a^b \equiv c \pmod{m}$$

This operation is fundamental in public-key cryptography algorithms like RSA (Rivest-Shamir-Adleman).

5.1.4   Example of Modular Exponentiation in RSA

Consider an example where Alice wants to send a confidential message to Bob using RSA. Bob has a public key (e, n), where e is the public exponent, and n is the modulus. Alice encrypts her message M using Bob's public key:

$$C \equiv M^e \pmod{n}$$

Bob, who possesses the corresponding private key, can then decrypt the message:

$$M \equiv C^d \pmod{n}$$

Here, d is the private exponent. The security of RSA relies on the difficulty of factoring the product of two large prime numbers, which forms the modulus n.

5.1.5   Modular Multiplication in Cryptography

Modular multiplication is another critical operation in cryptographic algorithms. When dealing with large numbers, modular multiplication helps prevent overflow and ensures the results remain within a specific range.

5.1.6 Example of Modular Multiplication in Cryptography

Consider a scenario where a cryptographic hash function uses modular multiplication. Let $H(x)$ be the hash function, and a and b are inputs:

$$H(x) \equiv (a \cdot b) \mod p$$

Here, p is a large prime number. The use of modular multiplication helps maintain the integrity of the hash function and ensures that the output is within the desired range.

5.1.7 Congruences in Cryptography

Congruences play a crucial role in cryptographic protocols. They are often used to represent equality relationships in modular arithmetic and are fundamental to the design of secure cryptographic algorithms.

5.1.8 Example of Congruences in Cryptography

Consider a scenario where Alice and Bob want to establish a shared secret key using the Diffie-Hellman key exchange. They choose a large prime p and a primitive root g. Each party selects a private key (a for Alice, b for Bob) and exchanges public keys:

$$A \equiv g^a \pmod{p}$$
$$B \equiv g^b \pmod{p}$$

They can then compute the shared secret key using their private keys and the received public keys:

$$\text{Shared Key} \equiv (g^a)^b \equiv (g^b)^a \pmod{p}$$

The security of the Diffie-Hellman key exchange relies on the difficulty of the discrete logarithm problem.

5.1.9   Chinese Remainder Theorem in Cryptography

The Chinese Remainder Theorem (CRT) is employed in various cryptographic systems, providing a method to speed up computations. It allows for the efficient handling of modular equations with different moduli.

5.1.10   Example of Chinese Remainder Theorem in Cryptography

Suppose a cryptographic system involves modular equations with different moduli, such as $x \equiv a \pmod{m_1}$ and $x \equiv b \pmod{m_2}$. The Chinese Remainder Theorem allows for the calculation of x using the individual congruences and their corresponding moduli.

$$x \equiv (a \cdot M_2 \cdot M_2^{-1}) + (b \cdot M_1 \cdot M_1^{-1}) \pmod{m_1 \cdot m_2}$$

Here, M_1 and M_2 are the product of the other moduli, and M_2^{-1} and M_1^{-1} are the modular inverses. The Chinese Remainder Theorem is particularly beneficial in optimizing cryptographic algorithms.

5.1.11   Applications of Cryptography in Real-world Scenarios

Cryptography finds widespread use in various real-world scenarios, including securing communication over the internet, protecting sensitive information in financial transactions, ensuring the integrity of data storage, and enabling secure access control systems.

5.1.12   Example: Secure Online Transactions

In the context of online transactions, cryptographic protocols, such as SSL/TLS, use a combination of symmetric and asymmetric encryption to secure communication between clients and servers. Public-key cryptography ensures the confidentiality of sensitive information, and symmetric-key cryptography provides efficient data transmission.

5.1.13 Example: Digital Signatures

Digital signatures, a crucial component of cryptographic systems, use mathematical algorithms based on modular arithmetic. A digital signature provides a way to verify the authenticity and integrity of a message or document. The signer uses their private key to generate the signature, and others can verify it using the corresponding public key.

5.1.14 Example: Password Hashing

In secure password storage, cryptographic hash functions employ modular arithmetic to convert passwords into hash values. The use of modular arithmetic helps create a one-way function, making it computationally infeasible to reverse the process and obtain the original password.

5.1.15 Challenges and Advancements in Cryptography

While cryptography has evolved to address emerging challenges, including quantum computing threats, advancements in mathematical techniques and cryptographic algorithms continue to enhance the security of information systems. The development of post-quantum cryptography, which considers algorithms secure against quantum attacks, is a significant area of research.

5.2 Number Theory

Number theory, a branch of mathematics that explores the properties and relationships of integers, plays a crucial role in understanding and leveraging modular arithmetic. In this section, we delve into the applications of number theory within the context of modular arithmetic, unraveling fundamental concepts, mathematical formulas, and practical examples.

5.2.1   Basic Concepts in Number Theory

Number theory encompasses a wide array of topics, including prime numbers, divisibility, congruences, and modular arithmetic. These concepts form the foundation for various applications in both pure mathematics and real-world problem-solving.

5.2.2   Primes and Primality Testing

Prime numbers, integers greater than 1 that have no positive divisors other than 1 and themselves, are central to number theory. Modular arithmetic often involves primes, and algorithms for primality testing are crucial in cryptographic applications.

5.2.3   Example: Fermat's Little Theorem

Fermat's Little Theorem, a fundamental result in number theory, states that if p is a prime number and a is an integer not divisible by p, then $a^{p-1} \equiv 1 \pmod{p}$. This theorem has applications in modular exponentiation and primality testing algorithms.

5.2.4   Divisibility and Modular Arithmetic

Divisibility rules, such as those for determining whether a number is divisible by 2, 3, 4, etc., are closely connected to modular arithmetic. The remainders of numbers upon division by small primes often reveal valuable information.

5.2.5   Example: Divisibility by 3 Rule

A number is divisible by 3 if and only if the sum of its digits is divisible by 3. In modular arithmetic terms, if $N = a_n a_{n-1} \ldots a_1 a_0$ is a decimal representation of a number, then $N \equiv a_n + a_{n-1} + \ldots + a_1 + a_0 \pmod{3}$.

5.2.6 Congruences in Number Theory

Congruences, expressions of the form $a \equiv b \pmod{m}$, are central to number theory. They describe relationships between integers with respect to a modulus m and find applications in solving diophantine equations and modular arithmetic algorithms.

5.2.7 Example: Linear Congruences

Consider the linear congruence $ax \equiv b \pmod{m}$. Solving this congruence involves finding an integer x such that ax leaves the same remainder as b upon division by m. This concept is foundational in modular arithmetic algorithms.

5.2.8 Chinese Remainder Theorem in Number Theory

The Chinese Remainder Theorem (CRT) is a powerful tool in number theory, providing a systematic way to solve a system of simultaneous congruences. It has applications in diverse areas, including cryptography, coding theory, and solving systems of linear diophantine equations.

5.2.9 Example: CRT Application in Diophantine Equations

Suppose we want to find an integer x such that $x \equiv 2 \pmod{3}$ and $x \equiv 3 \pmod{4}$. The Chinese Remainder Theorem allows us to solve this system and find a unique solution modulo 12.

5.2.10 Number Theory in Cryptography

Number theory forms the backbone of many cryptographic algorithms. The difficulty of certain number-theoretic problems, such as factoring large numbers and computing discrete logarithms, underpins the security of widely-used cryptographic protocols.

5.2.11   Example: RSA Algorithm

The RSA algorithm, a cornerstone of modern cryptography, relies on the difficulty of factoring the product of two large prime numbers. The security of RSA is grounded in the computational complexity of certain number-theoretic problems.

5.2.12   Applications in Coding Theory

Number theory finds applications in coding theory, particularly in designing error-detecting and error-correcting codes. Concepts like finite fields and algebraic structures play a vital role in constructing efficient codes.

5.2.13   Example: Reed-Solomon Codes

Reed-Solomon codes, widely used in data storage and communication systems, are constructed based on polynomials over finite fields. The algebraic properties of finite fields, rooted in number theory, contribute to the effectiveness of Reed-Solomon codes.

5.2.14   Number Theory in Cryptanalysis

Understanding number theory is crucial not only for designing secure cryptographic systems but also for cryptanalysis—the art of breaking cryptographic systems. Analyzing the mathematical structure of cryptographic algorithms often involves intricate number-theoretic concepts.

5.2.15   Example: Pollard's Rho Algorithm

Pollard's Rho algorithm, a probabilistic algorithm for integer factorization, leverages number-theoretic properties to efficiently find non-trivial factors of composite numbers. The algorithm's success relies on the distribution of primes.

5.2.16   Applications in Random Number Generation

Number theory plays a role in the generation of pseudorandom numbers, which find applications in simulations, cryptography, and statistical sampling. Certain number-theoretic properties are exploited to ensure the unpredictability and uniform distribution of generated numbers.

5.2.17   Example: Linear Congruential Generator

Linear congruential generators use modular arithmetic to produce sequences of pseudorandom numbers. The choice of parameters in such generators is influenced by number-theoretic considerations to achieve desirable statistical properties.

5.2.18   Connections to Algebraic Structures

Number theory exhibits close connections to various algebraic structures, including groups, rings, and fields. The study of these structures enriches the understanding of number-theoretic concepts and their applications.

5.2.19   Example: Finite Fields in Coding Theory

Finite fields, algebraic structures characterized by a finite number of elements, play a pivotal role in coding theory. The algebraic properties of finite fields, explored in number theory, contribute to the design of efficient error-correcting codes.

5.2.20   Research Frontiers in Number Theory

Ongoing research in number theory explores new frontiers, including the distribution of prime numbers, arithmetic algebraic geometry, and the study of elliptic curves. Advances in these areas have implications for both theoretical mathematics and practical applications.

5.2.21   Example: Birch and Swinnerton-Dyer Conjecture

The Birch and Swinnerton-Dyer conjecture, a prominent open problem in number theory, posits a deep connection between the algebraic structure of elliptic curves and the distribution of rational points on these curves. A resolution of this conjecture could have profound implications for both number theory and cryptography.

5.3   Computer Science

The integration of modular arithmetic into computer science is fundamental, influencing algorithms, data structures, cryptography, and various other aspects. In this section, we explore the applications of modular arithmetic in computer science, elucidating key concepts, mathematical formulas, and practical examples.

5.3.1   Binary Representation and Modular Arithmetic

In computer science, modular arithmetic finds applications in binary representation and bitwise operations. The modulus operation is often employed to keep values within a specific range, particularly in scenarios where overflow can occur.

5.3.2   Example: Binary Addition with Modular Arithmetic

Consider the addition of two binary numbers 1011 and 1101, with a modulus of 2^4. Standard binary addition yields 11000, but applying modular arithmetic with a modulus of 2^4 results in 1000, emphasizing the importance of modular arithmetic in managing overflow.

5.3.3   Hash Functions and Modular Arithmetic

Hash functions, crucial in computer science for data integrity and indexing, often employ modular arithmetic to map large data sets to a fixed range. The modulus operation ensures that hash values fit within the allotted storage space.

5.3.4 Example: Hashing with Modular Arithmetic

A simple hash function might involve summing the ASCII values of characters in a string and taking the modulus M to obtain a hash code. For instance, hashing "abc" might involve $(97 + 98 + 99) \mod M$.

5.3.5 Modular Exponentiation in Cryptography

In cryptographic algorithms, modular exponentiation plays a central role. Efficient algorithms for modular exponentiation, such as the square-and-multiply method, contribute to the security and performance of cryptographic systems.

5.3.6 Example: Modular Exponentiation in RSA

In the RSA algorithm, modular exponentiation is utilized during both encryption and decryption. If Alice wants to send a confidential message to Bob, she raises the message to the power of Bob's public key exponent modulo the public modulus.

5.3.7 Modular Arithmetic in Data Structures

Modular arithmetic finds applications in data structures, particularly in circular buffers and addressing within fixed-size arrays. It facilitates efficient implementation and manipulation of data in memory.

5.3.8 Example: Circular Buffer Indexing

A circular buffer with N elements often uses modular arithmetic for indexing. When incrementing the buffer index, taking the modulus N ensures that the index wraps around when reaching the buffer's size, creating a circular behavior.

5.3.9 Error Detection and Correction Codes

Modular arithmetic is integral to the design of error-detecting and error-correcting codes. Techniques from number theory, such as cyclic redundancy checks (CRC), leverage modular arithmetic to provide robust error-handling capabilities.

5.3.10 Example: CRC Checksum Calculation

In CRC, a polynomial is divided by another polynomial using binary polynomial division. The remainder, obtained through modular arithmetic, serves as the checksum that can detect errors in transmitted data.

5.3.11 Random Number Generation in Algorithms

Algorithms often require the generation of random numbers. Modular arithmetic helps create pseudorandom number generators, ensuring that the generated sequence exhibits statistical properties and cycles at sufficiently long intervals.

5.3.12 Example: Linear Congruential Generator

Linear congruential generators utilize modular arithmetic to generate pseudo-random numbers. The recurrence relation $X_{n+1} = (aX_n + c) \mod m$ defines the next value in the sequence, where a, c, and m are carefully chosen constants.

5.3.13 Graph Algorithms and Modular Arithmetic

In graph algorithms, modular arithmetic can be applied to problems involving cycles and connectivity. Techniques like the Union-Find data structure utilize modular arithmetic to efficiently manage disjoint sets.

5.3.14 Example: Union-Find with Path Compression

In Union-Find, the path compression step involves collapsing the structure of disjoint sets. Modular arithmetic helps maintain the integrity of the data structure by ensuring that each element points to a unique representative.

5.3.15 Modular Arithmetic in Computer Graphics

Computer graphics often involves operations on pixel values and colors, where modular arithmetic ensures that color values remain within the valid range. This prevents artifacts such as overflow or underflow.

5.3.16 Example: Color Blending with Modular Arithmetic

When blending colors, the resulting intensity for each color channel can be computed using modular arithmetic. For instance, adding two color values modulo 256 ensures that the result stays within the valid range for an 8-bit color channel.

5.3.17 Number Representation in Computer Architecture

In computer architecture, modular arithmetic is relevant to the representation of numbers using finite word sizes. Modular reduction is employed to fit results within the available bit width.

5.3.18 Example: Modular Reduction in Arithmetic Logic Unit (ALU)

In an ALU, modular reduction ensures that the output of arithmetic operations fits within the specified word size. For example, adding two n-bit numbers may require modular reduction to n bits.

5.3.19 Efficient Algorithms with Modular Arithmetic

Algorithms in computer science often exploit modular arithmetic for efficiency. The reduction of intermediate results modulo a chosen modulus can lead to faster computations and reduced memory requirements.

5.3.20   Example: Matrix Exponentiation with Modular Arithmetic

In certain algorithms, such as those involving matrix exponentiation, modular arithmetic allows for the reduction of intermediate results. This is particularly valuable in scenarios where numbers become extremely large.

5.3.21   Challenges and Opportunities in Computer Science

While modular arithmetic is a powerful tool in computer science, challenges exist, particularly in cryptographic applications with emerging threats like quantum computing. Researchers explore post-quantum cryptography and novel algorithmic techniques.

5.3.22   Example: Quantum-resistant Cryptography

The development of cryptographic algorithms resistant to quantum attacks is a pressing concern. Lattice-based cryptography, a field rooted in modular arithmetic, is explored for its potential to withstand quantum threats.

Chapter 6

Advanced Topics

6.1 Chinese Remainder Theorem

The Chinese Remainder Theorem (CRT) is a powerful result in number theory with widespread applications in modular arithmetic. In this section, we delve into the Chinese Remainder Theorem, exploring its mathematical formulation, algorithms for computation, and practical examples of its usage.

6.1.1 Mathematical Formulation of the CRT

The Chinese Remainder Theorem addresses systems of simultaneous linear congruences. Given a set of equations of the form:

$$x \equiv a_1 \pmod{m_1}$$

$$x \equiv a_2 \pmod{m_2}$$

$$\vdots$$

$$x \equiv a_k \pmod{m_k}$$

where $m_1, m_2, \ldots, m_k$ are pairwise coprime (i.e., their greatest common divisors are all 1), the CRT asserts the existence of a unique solution x modulo $M = m_1 \cdot m_2 \cdot \ldots \cdot m_k$.

6.1.2   Chinese Remainder Theorem - Existence and Uniqueness

The existence and uniqueness of a solution modulo M in the CRT are guaranteed by the pairwise coprimality of the moduli. This ensures that the Chinese Remainder Theorem provides a one-to-one correspondence between the residue classes and the solution modulo M.

6.1.3   Example: Solving a System of Congruences

Consider the system:

$$x \equiv 2 \pmod 3$$
$$x \equiv 3 \pmod 5$$
$$x \equiv 2 \pmod 7$$

The moduli 3, 5, and 7 are pairwise coprime. Using the CRT, we find a unique solution modulo $M = 3 \cdot 5 \cdot 7 = 105$. The solution is $x \equiv 23 \pmod{105}$.

6.1.4   Algorithm for Computing the CRT

Several algorithms exist for computing the solution using the CRT. One commonly used method is the Chinese Remainder Theorem algorithm, which involves computing partial remainders and combining them to obtain the final solution.

6.1.5   Chinese Remainder Theorem Algorithm

Given a system of congruences $x \equiv a_i \pmod{m_i}$ for $i = 1, 2, \ldots, k$, the Chinese Remainder Theorem algorithm involves the following steps:

1. Compute $M = m_1 \cdot m_2 \cdot \ldots \cdot m_k$. 2. For each i, compute $M_i = M/m_i$. 3. Find the modular inverse N_i of M_i modulo m_i using the Extended Euclidean Algorithm. 4. Compute $y_i = M_i \cdot N_i$ for each i. 5. The solution x is given by $x = (a_1 \cdot y_1 + a_2 \cdot y_2 + \ldots + a_k \cdot y_k) \mod M$.

6.1.6   Example: Applying the CRT Algorithm

Let's use the CRT algorithm to solve the system:

$$x \equiv 2 \pmod 3$$

$$x \equiv 3 \pmod 5$$

$$x \equiv 2 \pmod 7$$

Following the algorithm, we find $M = 105$, $M_1 = 35$, $M_2 = 21$, $M_3 = 15$. The modular inverses are $N_1 = 23$, $N_2 = 1$, $N_3 = 13$. Finally, $y_1 = 2$, $y_2 = 15$, $y_3 = 26$. The solution is $x = (2 \cdot 2 + 3 \cdot 15 + 2 \cdot 26) \mod 105 = 23$.

6.1.7   Applications of the Chinese Remainder Theorem

The Chinese Remainder Theorem has diverse applications, ranging from number theory to computer science and cryptography. Its efficiency in solving systems of congruences with coprime moduli makes it a valuable tool.

6.1.8   Number Theory Applications

In number theory, the CRT is employed to simplify calculations involving large integers and modular arithmetic. It provides an elegant solution to problems related to congruences and divisibility.

6.1.9   Example: Divisibility by Primes

Suppose we want to check whether a large number N is divisible by both 3 and 5. Using the CRT, we can check the remainders of N modulo 3 and 5 separately, making the overall divisibility check more manageable.

6.1.10   Computer Science and Cryptography Applications

The CRT finds applications in computer science, especially in optimization problems involving modular arithmetic. In cryptography, the CRT enhances the efficiency of certain algorithms, such as RSA.

6.1.11   Example: RSA Key Generation

In RSA cryptography, key generation involves choosing two large primes p and q, computing $N = p \cdot q$, and finding the public and private exponents. The CRT facilitates the efficient calculation of the private exponent.

6.1.12   Coding Theory Applications

In coding theory, the CRT is utilized in the construction of error-correcting codes. Techniques involving the CRT contribute to the development of efficient coding schemes.

6.1.13   Example: Reed-Solomon Codes

Reed-Solomon codes, widely used in data storage and transmission, utilize the CRT in decoding procedures. The CRT aids in efficiently recovering information from received codewords.

6.1.14   Research Frontiers in the CRT

Ongoing research explores extensions and generalizations of the CRT, including applications in algebraic geometry and algebraic number theory. The development of algorithms and techniques for handling non-coprime moduli is an active area of investigation.

6.1.15   Example: CRT for Non-coprime Moduli

Extensions of the CRT for systems with non-coprime moduli involve addressing challenges arising from common factors. Research aims to broaden the applicability of the CRT to a wider range of scenarios.

6.2 Fermat's Little Theorem

Fermat's Little Theorem is a fundamental result in number theory with significant applications in modular arithmetic. In this section, we explore the mathematical formulation of Fermat's Little Theorem, discuss its implications, and provide examples of its applications.

6.2.1 Mathematical Formulation of Fermat's Little Theorem

Fermat's Little Theorem states that for any prime number p and any integer a not divisible by p, the following congruence holds:

$$a^{p-1} \equiv 1 \pmod{p}$$

This theorem provides a powerful tool for analyzing the properties of integers in modular arithmetic.

6.2.2 Implications and Applications

Fermat's Little Theorem has diverse applications, ranging from primality testing to the design of cryptographic algorithms. Its simplicity and efficiency make it a valuable tool in number theory and related fields.

6.2.3 Example: Primality Testing

One application of Fermat's Little Theorem is in probabilistic primality testing. Given an odd prime p, if $a^{p-1} \not\equiv 1 \pmod{p}$ for some randomly chosen a, then p is composite. While the test is probabilistic, its efficiency makes it useful in practice.

6.2.4 Carmichael Numbers and Fermat's Little Theorem

Carmichael numbers are composite numbers that satisfy Fermat's Little Theorem for all values of a coprime to the number. Exploring the properties of

Carmichael numbers provides insights into the limitations of Fermat's Little Theorem in certain contexts.

6.2.5   Example: Carmichael Number

Consider the Carmichael number 561. For any a coprime to 561, $a^{560} \equiv 1$ (mod 561). While 561 is composite, it satisfies Fermat's Little Theorem.

6.2.6   Applications in Cryptography

Fermat's Little Theorem is utilized in the field of cryptography, particularly in the RSA algorithm. The theorem plays a crucial role in ensuring the security of the algorithm.

6.2.7   Example: RSA Encryption

In RSA encryption, the security relies on the difficulty of factoring large composite numbers. Fermat's Little Theorem is involved in the selection of suitable parameters to make the factorization process challenging for potential adversaries.

6.2.8   Fermat Pseudoprimes

Fermat pseudoprimes are composite numbers that satisfy Fermat's Little Theorem for all a coprime to the number. Understanding these pseudoprimes contributes to the analysis of Fermat's Little Theorem and its limitations.

6.2.9   Example: Fermat Pseudoprime

The number 341 is a Fermat pseudoprime, as $2^{340} \equiv 1$ (mod 341). Despite being composite, it satisfies Fermat's Little Theorem for all a coprime to 341.

6.2.10 Extensions and Generalizations

While Fermat's Little Theorem is stated for prime moduli, various extensions and generalizations exist for composite moduli. These generalizations involve additional conditions or constraints on the integers involved.

6.2.11 Example: Euler's Totient Theorem

Euler's Totient Theorem is a generalization of Fermat's Little Theorem, applicable to composite moduli. It states that for any integer a coprime to n, where n is a positive integer, the congruence $a^{\phi(n)} \equiv 1 \pmod{n}$ holds, where $\phi(n)$ is Euler's totient function.

6.2.12 Research Frontiers in Fermat's Little Theorem

Ongoing research explores variations and refinements of Fermat's Little Theorem, especially in the context of composite numbers. Theoretical advancements contribute to a deeper understanding of the theorem's properties and limitations.

6.2.13 Example: Frobenius Pseudoprimes

Frobenius pseudoprimes are composite numbers that satisfy a generalized form of Fermat's Little Theorem. Research investigates the distribution and properties of these pseudoprimes, adding nuance to the understanding of Fermat's Little Theorem.

6.2.14 Applications in Coding Theory

Fermat's Little Theorem finds applications in coding theory, particularly in the design and analysis of error-correcting codes. The properties of the theorem contribute to the efficiency and reliability of certain coding schemes.

6.2.15   Example: Reed-Solomon Codes

Reed-Solomon codes, widely used in data storage and communication, leverage properties akin to Fermat's Little Theorem. The theorem's insights aid in designing robust error-correcting codes.

6.3   Euler's Totient Function

Euler's Totient Function, denoted by $\phi(n)$, is a crucial concept in number theory with diverse applications in modular arithmetic. In this section, we explore the mathematical definition of Euler's Totient Function, discuss its properties, and provide examples of its applications.

6.3.1   Mathematical Definition of Euler's Totient Function

Euler's Totient Function $\phi(n)$ is defined as the count of positive integers less than or equal to n that are coprime to n. In mathematical terms:

$$\phi(n) = \mathrm{Card}\{k \mid 1 \leq k \leq n,\ \gcd(k, n) = 1\}$$

The function counts the number of integers in the range $[1, n]$ that share no common factors with n other than 1.

6.3.2   Properties of Euler's Totient Function

Euler's Totient Function exhibits several important properties:

1. For prime numbers p, $\phi(p) = p - 1$. 2. For any positive integer n, $\sum_{d|n} \phi(d) = n$, where the sum is taken over all positive divisors d of n. 3. If $n = p_1^{a_1} p_2^{a_2} \ldots p_k^{a_k}$ is the prime factorization of n,

then $\phi(n) = n \left(1 - \frac{1}{p_1}\right)\left(1 - \frac{1}{p_2}\right)\ldots\left(1 - \frac{1}{p_k}\right)$.

Example: Euler's Totient Function for $n = 12$

Let's calculate $\phi(12)$. The prime factorization of 12 is $2^2 \cdot 3^1$. Using the formula, we get:

$$\phi(12) = 12 \left(1 - \frac{1}{2}\right) \left(1 - \frac{1}{3}\right) = 4 \cdot \frac{1}{2} \cdot \frac{2}{3} = 4$$

So, $\phi(12) = 4$, indicating that there are four positive integers less than or equal to 12 that are coprime to 12.

6.3.3 Applications in Number Theory

Euler's Totient Function finds applications in various number theory problems, particularly those related to congruences and primitive roots. The function is a key tool in analyzing the distribution of coprime integers.

6.3.4 Example: Fermat's Little Theorem Revisited

Fermat's Little Theorem can be expressed in terms of Euler's Totient Function. For any prime p and integer a not divisible by p, Fermat's Little Theorem can be stated as $a^{\phi(p)} \equiv 1 \pmod{p}$.

6.3.5 Applications in Cryptography

In cryptography, Euler's Totient Function plays a crucial role in the RSA algorithm. The function aids in the selection of suitable parameters for ensuring the security of the algorithm.

6.3.6 Example: RSA Key Generation

In RSA cryptography, key generation involves choosing two large primes p and q, computing $n = pq$, and finding the public and private exponents. Euler's Totient Function is used to calculate $\phi(n) = (p-1)(q-1)$, influencing the selection of the public and private exponents.

6.3.7 Applications in Group Theory

Euler's Totient Function is linked to group theory through the concept of primitive roots. The function helps determine the existence and count of primitive roots modulo a given integer.

6.3.8   Example: Primitive Roots

A primitive root modulo n is an integer g such that the powers of g generate all coprime residues modulo n. Euler's Totient Function is instrumental in determining whether primitive roots exist for a given modulus.

6.3.9   Research Frontiers in Euler's Totient Function

Ongoing research explores extensions and generalizations of Euler's Totient Function, particularly in the context of composite moduli. The study of totient functions for composite numbers involves additional complexities and considerations.

6.3.10   Example: Generalized Totient Functions

Generalized totient functions extend the concept of Euler's Totient Function to composite moduli. These functions capture the coprimality relationships within the set of residues modulo a composite number.

6.3.11   Applications in Coding Theory

Euler's Totient Function contributes to the design and analysis of error-correcting codes in coding theory. The function's properties are leveraged to ensure the efficiency and reliability of certain coding schemes.

6.3.12   Example: Coding Theory Applications

In coding theory, the properties of Euler's Totient Function are utilized in constructing codes with desirable properties, such as low cross-correlation and good error-correcting capabilities.

Chapter 7

Modular Linear Equations

7.1 Solution Methods

In this section, we explore various solution methods for modular linear equations, which are fundamental in modular arithmetic. Modular linear equations are expressions of the form $ax \equiv b \pmod{m}$, where a, b, and m are integers and $m > 0$. We delve into different techniques for finding solutions to such equations.

7.1.1 Basic Concepts

To understand solution methods, let's revisit the basic concepts of modular arithmetic. In a modular congruence $a \equiv b \pmod{m}$, a and b have the same remainder when divided by m. In modular linear equations, our goal is to find values of x that satisfy the congruence $ax \equiv b \pmod{m}$.

7.1.2 Direct Solution

One straightforward method is direct solution, where we isolate x by multiplying both sides of the equation by the modular inverse of a (if it exists). The modular inverse is the number a^{-1} such that $aa^{-1} \equiv 1 \pmod{m}$. If a and m are coprime, the modular inverse exists.

7.1.3   Example: Direct Solution

Consider the equation $3x \equiv 4 \pmod 7$. Since 3 and 7 are coprime, the modular inverse $3^{-1} \equiv 5 \pmod 7$. Multiplying both sides by 5, we get $x \equiv 6 \pmod 7$.

7.1.4   Extended Euclidean Algorithm

For cases where the modular inverse is not immediately obvious, the Extended Euclidean Algorithm can be employed to find it. This algorithm calculates the greatest common divisor (GCD) of a and m along with coefficients x' and y' such that $ax' + my' = \mathrm{GCD}(a, m)$.

7.1.5   Example: Extended Euclidean Algorithm

Let's find the solution for $9x \equiv 2 \pmod{13}$. Using the Extended Euclidean Algorithm, we find $9^{-1} \equiv 3 \pmod{13}$. Multiplying both sides by 3, we obtain $x \equiv 11 \pmod{13}$.

7.1.6   Chinese Remainder Theorem

For systems of modular linear equations, the Chinese Remainder Theorem (CRT) provides a powerful solution method. The CRT is particularly useful when dealing with congruences modulo pairwise coprime moduli.

7.1.7   Example: Chinese Remainder Theorem

Consider the system:

$$\begin{cases} 3x \equiv 2 \pmod 5 \\ 4x \equiv 3 \pmod 7 \end{cases}$$

Using the CRT, we find $x \equiv 13 \pmod{35}$ as the simultaneous solution.

7.1.8 Systematic Enumeration

In certain cases, especially with small moduli, systematic enumeration of potential solutions can be a viable method. This involves trying all possible values of x within a given range until a solution is found.

7.1.9 Example: Systematic Enumeration

For $5x \equiv 3 \pmod{11}$, systematic enumeration involves checking values of x within the range $[0, 10]$. The solution is $x \equiv 7 \pmod{11}$.

7.1.10 Applications in Cryptography

The solution methods for modular linear equations find extensive applications in cryptography, particularly in the implementation of public-key cryptosystems like RSA. The ability to efficiently solve modular linear equations is crucial for cryptographic protocols.

7.1.11 Example: RSA Key Generation

In RSA key generation, modular linear equations play a central role. The selection of suitable values for p, q, and the public exponent involves solving modular linear equations to ensure the security of the cryptographic system.

7.1.12 Error Detection and Correction

Modular arithmetic, including the solution of modular linear equations, is employed in error-detection and correction codes. The ability to solve equations modulo a prime or composite number contributes to the reliability of such coding schemes.

7.1.13 Example: Coding Theory

In coding theory, the solution of modular linear equations is utilized in the design of error-correcting codes. The properties of modular arithmetic aid in constructing codes with desirable characteristics.

7.1.14   Research Frontiers

Ongoing research explores advanced techniques for solving modular linear equations efficiently, especially in scenarios involving large primes and complex systems. The development of algorithms for specific cases remains an active area of investigation.

7.1.15   Example: Elliptic Curve Cryptography

Elliptic Curve Cryptography relies on solving equations involving elliptic curves defined over finite fields. Efficient algorithms for solving modular equations in this context contribute to the security of modern cryptographic systems.

7.2   Linear Diophantine Equations

In this section, we delve into the realm of Linear Diophantine Equations within the context of modular arithmetic. A Linear Diophantine Equation is an equation of the form $ax + by = c$, where a, b, c, and the variables x and y are integers. We explore methods for finding solutions to such equations, with a focus on their applications in modular arithmetic.

7.2.1   Basic Concepts

A Linear Diophantine Equation can be expressed in modular form as $ax \equiv c$ $(\bmod\ m)$ if and only if $\gcd(a, m) \mid c$. The existence of solutions depends on whether the greatest common divisor (gcd) of a and m divides c.

7.2.2   Direct Solution

One method for solving Linear Diophantine Equations involves finding a particular solution and then generating the general solution. This often requires the use of the Extended Euclidean Algorithm to find a particular solution.

7.2.3   Example: Direct Solution

Consider the equation $21x+14y = 35$. Using the Extended Euclidean Algorithm, a particular solution is $x = -1, y = 2$. The general solution is then given by $x = -1 + 14t$ and $y = 2 - 21t$ for any integer t.

7.2.4   Parametric Representations

Linear Diophantine Equations often have infinitely many solutions, and these solutions can be represented parametrically using parameters or variables. The parameters allow us to express all solutions in a concise form.

7.2.5   Example: Parametric Representation

For the equation $15x + 10y = 25$, a particular solution is $x = 1, y = 1$. The parametric solution is then given by $x = 1 + 2t$ and $y = 1 - 3t$ for any integer t.

7.2.6   Chinese Remainder Theorem

The Chinese Remainder Theorem (CRT) can be applied to solve systems of Linear Diophantine Equations efficiently. This method is particularly useful when dealing with congruences modulo pairwise coprime moduli.

7.2.7   Example: CRT for Linear Diophantine Equations

Consider the system:
$$\begin{cases} 3x + 5y \equiv 1 \pmod 7 \\ 2x - y \equiv 3 \pmod 5 \end{cases}$$

Using CRT, we find $x \equiv 2 \pmod 7$ and $y \equiv 4 \pmod 5$ as the simultaneous solution.

7.2.8   Applications in Number Theory

Linear Diophantine Equations have applications in number theory, particularly in the study of Diophantine problems. These equations are often encountered in problems related to divisibility and number theory.

7.2.9   Example: Frobenius Coin Problem

The Frobenius Coin Problem involves finding the largest value that cannot be expressed as a linear combination of given integers. This problem can be formulated as a Linear Diophantine Equation and is essential in number theory.

7.2.10   Applications in Cryptography

Linear Diophantine Equations are foundational in cryptography, especially in the analysis of cryptographic algorithms. These equations play a role in the study of security properties and key generation.

7.2.11   Example: RSA Key Generation

In RSA key generation, Linear Diophantine Equations are involved in determining suitable values for the public and private keys. The parameters of the RSA algorithm are chosen to satisfy specific Diophantine conditions.

7.2.12   Error Detection and Correction

Linear Diophantine Equations find applications in error-detection and correction codes. The solutions to these equations contribute to the design and analysis of error-correcting codes.

7.2.13   Example: Linear Diophantine Equations in Coding Theory

In coding theory, Linear Diophantine Equations are used to design codes with desirable properties. The solutions to these equations influence the selection of parameters for error-correcting codes.

7.2.14   Research Frontiers

Ongoing research explores advanced techniques for solving Linear Diophantine Equations efficiently, especially in scenarios involving large integers and complex systems. The development of algorithms for specific cases remains an active area of investigation.

7.2.15   Example: Elliptic Curve Cryptography

Linear Diophantine Equations are fundamental in Elliptic Curve Cryptography, where solutions to such equations over elliptic curves are employed in cryptographic protocols. The study of efficient algorithms in this context is crucial for the security of cryptographic systems.

Chapter 8

Finite Fields

8.1 Definition

In this section, we explore the fundamental concept of finite fields within the context of modular arithmetic. A finite field, often denoted as $\mathbb{F}_q$, is a mathematical structure that exhibits both addition and multiplication operations, satisfying specific properties. We begin by defining finite fields and discussing their essential characteristics.

8.1.1 Basic Concepts

A finite field $\mathbb{F}_q$ consists of a finite set of elements, where q is a prime power ($q = p^n$, where p is a prime and n is a positive integer). The elements in the finite field are residues modulo p raised to powers less than n. The addition and multiplication operations in $\mathbb{F}_q$ are performed modulo p.

8.1.2 Mathematical Definition

The elements of a finite field $\mathbb{F}_q$ are often denoted as $\{0, 1, \ldots, q - 1\}$. The addition and multiplication operations are defined as follows:

- Addition: $a + b \equiv c \pmod{q}$

- Multiplication: $a \cdot b \equiv d \pmod{q}$

Here, a, b, c, d are elements in $\mathbb{F}_q$, and the operations are performed modulo q.

Example: Finite Field $\mathbb{F}_5$

Let's consider the finite field $\mathbb{F}_5$. The elements are $\{0, 1, 2, 3, 4\}$. Addition and multiplication are performed modulo 5. For instance, in $\mathbb{F}_5$:

- $2 + 3 \equiv 0 \pmod{5}$

- $4 \cdot 3 \equiv 2 \pmod{5}$

8.1.3   Field Properties

Finite fields exhibit several key properties that distinguish them from general rings or groups. These properties include closure under addition and multiplication, the existence of additive and multiplicative inverses, commutativity, and the distributive property.

8.1.4   Example: Closure Property

In $\mathbb{F}_7$, if we add any two elements or multiply any two elements, the result is always an element within the field. For instance, $4 + 5 \equiv 2 \pmod{7}$ and $3 \cdot 6 \equiv 3 \pmod{7}$.

8.1.5   Prime Fields and Extension Fields

A prime field is a finite field with $q = p$, where p is a prime number. Extension fields are finite fields obtained by extending the order of a smaller finite field.

Example: Prime Field $\mathbb{F}_3$

The finite field $\mathbb{F}_3$ is a prime field. It consists of elements $\{0, 1, 2\}$ with addition and multiplication performed modulo 3.

Example: Extension Field $\mathbb{F}_{2^3}$

The extension field $\mathbb{F}_{2^3}$ is an extension of $\mathbb{F}_2$. It consists of elements with 3-bit binary representations, and addition/multiplication are performed modulo a specific irreducible polynomial.

8.1.6 Applications in Coding Theory

Finite fields are extensively used in coding theory, particularly in the design and analysis of error-correcting codes. The algebraic properties of finite fields facilitate efficient encoding and decoding processes.

8.1.7 Example: Reed-Solomon Codes

Reed-Solomon codes, widely used in data storage and communication, leverage the properties of finite fields for error correction. The symbols of the code are elements of a finite field, and arithmetic operations are performed in the field.

8.1.8 Applications in Cryptography

Finite fields play a crucial role in modern cryptography, especially in public-key cryptosystems. The security of these systems relies on the algebraic properties of finite fields.

8.1.9 Example: Elliptic Curve Cryptography

Elliptic Curve Cryptography (ECC) involves finite fields. The group operations in ECC are defined over points on elliptic curves, and these points form finite fields. ECC provides secure cryptographic protocols.

8.1.10 Research Frontiers

Ongoing research explores advanced topics related to finite fields, including efficient algorithms for arithmetic operations, construction of specific finite fields, and their applications in emerging fields such as post-quantum cryptography.

8.1.11   Example: Efficient Finite Field Arithmetic

Efficient algorithms for arithmetic operations in finite fields, such as multiplication and exponentiation, are critical for the performance of cryptographic systems. Researchers continually seek improvements in these algorithms.

8.2   Operations in Finite Fields

In this section, we delve into the fundamental operations within finite fields, exploring the addition, subtraction, multiplication, and division of elements in a finite field $\mathbb{F}_q$. Understanding these operations is essential for various applications, including coding theory and cryptography.

8.2.1   Addition in Finite Fields

Addition in a finite field $\mathbb{F}_q$ is performed modulo q. Given two elements a, b in the field, the sum $a + b$ is calculated as $(a + b) \mod q$.

Example: Addition in $\mathbb{F}_7$

Consider $\mathbb{F}_7$ with elements $\{0, 1, 2, 3, 4, 5, 6\}$. The addition $3 + 5$ in $\mathbb{F}_7$ is $(3 + 5) \mod 7 = 1$.

8.2.2   Subtraction in Finite Fields

Subtraction is also performed modulo q in a finite field $\mathbb{F}_q$. For elements a, b, the difference $a - b$ is calculated as $(a - b) \mod q$.

Example: Subtraction in $\mathbb{F}_{11}$

Consider $\mathbb{F}_{11}$ with elements $\{0, 1, 2, 3, 4, 5, 6, 7, 8, 9, 10\}$. The subtraction $8 - 3$ in $\mathbb{F}_{11}$ is $(8 - 3) \mod 11 = 5$.

8.2.3   Multiplication in Finite Fields

Multiplication in a finite field $\mathbb{F}_q$ is also performed modulo q. Given elements a, b, the product $a \cdot b$ is calculated as $(a \cdot b) \mod q$.

Example: Multiplication in $\mathbb{F}_{13}$

Consider $\mathbb{F}_{13}$ with elements $\{0, 1, 2, 3, 4, 5, 6, 7, 8, 9, 10, 11, 12\}$. The multiplication $6 \cdot 9$ in $\mathbb{F}_{13}$ is $(6 \cdot 9) \mod 13 = 10$.

8.2.4 Division in Finite Fields

Division in a finite field $\mathbb{F}_q$ involves finding the multiplicative inverse. Given elements a, b where $b \neq 0$, the division a/b is equivalent to $a \cdot b^{-1}$, where b^{-1} is the multiplicative inverse of b modulo q.

Example: Division in $\mathbb{F}_{17}$

Consider $\mathbb{F}_{17}$ with elements $\{0, 1, 2, 3, 4, 5, 6, 7, 8, 9, 10, 11, 12, 13, 14, 15, 16\}$. The division $12/8$ in $\mathbb{F}_{17}$ is equivalent to $12 \cdot 15$ since $8^{-1} \equiv 15 \mod 17$, resulting in $(12 \cdot 15) \mod 17 = 9$.

8.2.5 Field Properties

Finite fields exhibit properties such as closure under operations, commutativity, associativity, and the existence of additive and multiplicative inverses. These properties make finite fields crucial in various mathematical applications.

Example: Field Properties in $\mathbb{F}_5$

Consider $\mathbb{F}_5$ with elements $\{0, 1, 2, 3, 4\}$. Addition and multiplication in $\mathbb{F}_5$ are closed operations, and each element has an additive and multiplicative inverse.

8.2.6 Applications in Coding Theory

Understanding operations in finite fields is essential in coding theory. Error-correcting codes, such as Reed-Solomon codes, rely on efficient arithmetic operations in finite fields for encoding and decoding processes.

8.2.7   Example: Reed-Solomon Code Multiplication

In Reed-Solomon codes, multiplication of symbols is performed in a finite field. The multiplication operation ensures the generation of codewords with desirable properties for error correction.

8.2.8   Applications in Cryptography

Finite fields play a crucial role in cryptographic protocols, especially in public-key cryptography. Operations in finite fields form the basis for secure cryptographic algorithms.

8.2.9   Example: Modular Exponentiation in RSA

In the RSA cryptosystem, modular exponentiation is a key operation performed in a finite field. The modulus operation ensures that computations are carried out within the finite field, maintaining the security of the algorithm.

8.2.10   Research Frontiers

Ongoing research explores advanced topics related to operations in finite fields, including the development of efficient algorithms for arithmetic operations and their application in emerging cryptographic protocols.

8.2.11   Example: Efficient Finite Field Arithmetic Algorithms

Efficient algorithms for operations in finite fields, such as fast multiplication and exponentiation techniques, are actively researched. These algorithms aim to improve the performance of cryptographic systems.

8.3 Applications

In this section, we explore diverse applications of finite fields, showcasing how these mathematical structures play a pivotal role in areas such as coding theory, cryptography, and error correction.

8.3.1 Coding Theory and Reed-Solomon Codes

Finite fields find extensive applications in coding theory, where error-correcting codes are designed to transmit information reliably over noisy channels. Reed-Solomon codes, in particular, leverage the algebraic properties of finite fields to achieve efficient error correction.

Consider a Reed-Solomon code over $\mathbb{F}_{11}$ with message symbols $m_0, m_1, \ldots, m_k$. The codeword $c(x)$ is generated by evaluating a polynomial over $\mathbb{F}_{11}$, where $c(x) = m_0 + m_1 x + \ldots + m_k x^k$.

8.3.2 Example: Reed-Solomon Code Encoding

Let $m(x) = 2 + 4x + 7x^2$ be a message polynomial. The corresponding Reed-Solomon codeword $c(x)$ is obtained by evaluating $m(x)$ in $\mathbb{F}_{11}$. The resulting codeword is $c(x) = 2 + 4x + 7x^2$.

8.3.3 Cryptography and Elliptic Curve Cryptography (ECC)

Finite fields play a central role in modern cryptography, and elliptic curve cryptography (ECC) is a prominent example. ECC involves operations on points defined over elliptic curves in finite fields, providing a foundation for secure cryptographic protocols.

Consider an elliptic curve defined over $\mathbb{F}_{17}$: $y^2 \equiv x^3 + 3x + 6 \mod 17$. Points on this curve form a finite field, and ECC operations are performed within this field.

8.3.4   Example: ECC Point Addition

Let $P = (3, 7)$ and $Q = (5, 1)$ be points on the elliptic curve. The sum $P + Q$ is calculated using ECC point addition formulas within $\mathbb{F}_{17}$.

8.3.5   Error Detection and Correction in Communication

In communication systems, particularly in data transmission and storage, error detection and correction are crucial. Finite fields enable the construction of powerful error-correcting codes that can detect and correct errors efficiently.

Consider a linear block code defined over $\mathbb{F}_8$ using a parity-check matrix. The code corrects errors by operating within the finite field and identifying the error patterns.

8.3.6   Example: Linear Block Code Decoding

For a received vector $\mathbf{r}$, the syndrome $\mathbf{s}$ is computed using the parity-check matrix. If $\mathbf{s}$ is non-zero, an error is detected, and the error location is identified using finite field operations.

8.3.7   Cryptographic Hash Functions

Cryptographic hash functions are fundamental in ensuring data integrity and security. Finite fields are employed in hash function design to achieve collision resistance and prevent unauthorized alterations.

Consider a hash function $H(x)$ designed over $\mathbb{F}_{2^{256}}$, where the input x is hashed to a 256-bit output using finite field operations.

8.3.8   Example: Hash Function Operation

For a given input x, the hash function $H(x)$ involves finite field operations such as polynomial modulo reduction to ensure the output remains within $\mathbb{F}_{2^{256}}$.

8.3.9 Cyclic Redundancy Check (CRC) Codes

Cyclic Redundancy Check (CRC) codes are widely used for error detection in network communications and storage systems. These codes, based on polynomial division over finite fields, efficiently detect errors in data.

Consider a CRC code defined over $\mathbb{F}_2$ with a generator polynomial. The transmitted data is appended with a CRC checksum obtained through polynomial division in $\mathbb{F}_2$.

8.3.10 Example: CRC Code Generation

Let the data be represented as a polynomial $D(x)$, and the generator polynomial be $G(x)$. The CRC checksum $R(x)$ is calculated as the remainder of $D(x) \cdot x^n \nabla \cdot G(x)$ within $\mathbb{F}_2$.

8.3.11 Digital Signatures and Public-Key Cryptography

Public-key cryptography relies on finite fields for secure digital signatures. Algorithms like the Digital Signature Algorithm (DSA) operate within finite fields to ensure the authenticity and integrity of digital signatures.

Consider the DSA algorithm over $\mathbb{F}_p$, where p is a large prime. The private key, public key, and signature generation involve finite field operations.

8.3.12 Example: DSA Signature Generation

Let (p, q, g) be the DSA parameters, x be the private key, and k be a random number. The DSA signature (r, s) is generated through finite field operations within $\mathbb{F}_p$.

8.3.13 Linear Feedback Shift Registers (LFSRs)

Linear Feedback Shift Registers are essential in pseudorandom sequence generation and stream ciphers. These registers, implemented using finite fields, exhibit linear feedback and efficient shift operations.

Consider an LFSR defined over $\mathbb{F}_2$ with feedback coefficients. The LFSR state is updated through bitwise XOR operations within $\mathbb{F}_2$.

8.3.14   Example: LFSR State Update

For an LFSR state $S = s_0 s_1 s_2$, the next state S' is obtained through feedback and shift operations, incorporating finite field arithmetic.

8.3.15   Research Frontiers

Ongoing research explores advanced applications of finite fields in emerging technologies. This includes the development of efficient algorithms for finite field operations and their integration into novel cryptographic protocols and error-correcting codes.

8.3.16   Example: Post-Quantum Cryptography

In the realm of post-quantum cryptography, researchers are exploring new cryptographic schemes based on the hardness of problems in finite fields. These schemes aim to resist attacks from quantum computers.

Chapter 9

Coding Theory

9.1 Linear Block Codes

Linear block codes are a fundamental concept in coding theory, providing a systematic way to encode messages for reliable communication over noisy channels. In this section, we delve into the definition, properties, and examples of linear block codes.

9.1.1 Definition of Linear Block Codes

A linear block code is a type of error-correcting code that operates on blocks of symbols. It is characterized by a linear relationship between the encoded message and the transmitted codeword. Mathematically, a linear block code of length n over a finite field $\mathbb{F}_q$ can be represented as:

$$C : \mathbb{F}_q^k \to \mathbb{F}_q^n$$

where k is the dimension of the code, representing the number of message symbols, and n is the length of the codeword.

9.1.2   Generator Matrix and Parity-Check Matrix

The key components of a linear block code are the generator matrix (G) and the parity-check matrix (H). The generator matrix is used to create codewords, while the parity-check matrix is employed for error detection and correction.

For an (n, k) linear block code, the generator matrix G has dimensions $k \times n$, and the parity-check matrix H has dimensions $(n - k) \times n$.

9.1.3   Example: (7, 4) Hamming Code

Consider the (7, 4) Hamming code over $\mathbb{F}_2$. The generator matrix G for this code is:

$$G = \begin{bmatrix} 1 & 0 & 0 & 0 & 1 & 1 & 0 \\ 0 & 1 & 0 & 0 & 0 & 1 & 1 \\ 0 & 0 & 1 & 0 & 1 & 1 & 1 \\ 0 & 0 & 0 & 1 & 1 & 0 & 1 \end{bmatrix}$$

This code can encode a 4-bit message into a 7-bit codeword using matrix multiplication.

9.1.4   Encoding Process

The encoding process for linear block codes involves multiplying the message vector (M) by the generator matrix (G) to obtain the codeword vector (C):

$$C = M \cdot G$$

9.1.5   Example: Encoding in the (7, 4) Hamming Code

Let $M = [1, 0, 1, 1]$ be a 4-bit message. The codeword C is obtained by multiplying M with the generator matrix G:

$$C = [1,0,1,1] \cdot \begin{bmatrix} 1 & 0 & 0 & 0 & 1 & 1 & 0 \\ 0 & 1 & 0 & 0 & 0 & 1 & 1 \\ 0 & 0 & 1 & 0 & 1 & 1 & 1 \\ 0 & 0 & 0 & 1 & 1 & 0 & 1 \end{bmatrix} = [1,0,1,1,0,1,0]$$

The codeword C is the transmitted sequence.

9.1.6 Decoding Process

Decoding in linear block codes involves the use of the parity-check matrix (H). The received vector (R) is multiplied by the transpose of the parity-check matrix (H^T) to check for errors.

If $H^T \cdot R^T = \mathbf{0}$, no errors are detected. Otherwise, the syndrome vector obtained is used to identify and correct errors.

9.1.7 Example: Decoding in the (7, 4) Hamming Code

Let $R = [1,1,1,1,0,1,0]$ be the received vector. The syndrome vector S is obtained by multiplying R with the transpose of the parity-check matrix H^T:

$$S = R \cdot H^T$$

If $S = \mathbf{0}$, no errors are detected. Otherwise, the position of the non-zero element in S indicates the position of the error, and correction can be applied.

9.1.8 Error Detection and Correction

Linear block codes are designed to detect and correct errors introduced during transmission. The minimum distance (d) of a code is a crucial parameter, representing the minimum number of symbol changes required to convert one valid codeword into another. A higher minimum distance results in better error-correcting capabilities.

9.1.9   Example: Minimum Distance in the (7, 4) Hamming Code

For the (7, 4) Hamming code, the minimum distance d is 3. This means the code can detect and correct up to 1 error. If two errors occur, the code may detect but cannot correct.

9.1.10   Performance Metrics

The performance of a linear block code is often evaluated using metrics such as the code rate (R), defined as the ratio of the message length (k) to the codeword length (n).

$$R = \frac{k}{n}$$

A higher code rate implies efficient use of bandwidth.

9.1.11   Example: Code Rate in the (7, 4) Hamming Code

For the (7, 4) Hamming code, the code rate is $R = \frac{4}{7}$.

9.1.12   Applications in Data Storage

Linear block codes find extensive use in data storage systems, where reliable retrieval of information is crucial. Examples include CDs, DVDs, and hard drives, where error correction is vital to ensure data integrity.

9.1.13   Example: CD Error Correction

Compact Discs (CDs) employ Reed-Solomon codes, a type of linear block code, for error correction. This ensures accurate playback of audio data even in the presence of scratches or other imperfections.

9.1.14   Applications in Telecommunication

In telecommunications, linear block codes are utilized to transmit data reliably over noisy communication channels. This is critical for maintaining the integrity of transmitted information.

9.1.15   Example: Satellite Communication

Satellite communication systems use error-correcting codes to mitigate the effects of signal interference and noise during data transmission. Linear block codes enhance the robustness of these systems.

9.1.16   Research Frontiers

Ongoing research in coding theory explores advanced topics related to linear block codes, including the design of codes with specific properties, the development of efficient decoding algorithms, and their application in emerging technologies such as 5G communication.

9.1.17   Example: Polar Codes

Polar codes, a class of linear block codes, have gained attention for their capacity-achieving properties. Researchers are investigating their application in various communication scenarios, including beyond-5G networks.

9.2   Cyclic Codes

Cyclic codes are a special class of linear block codes with unique algebraic properties. In this section, we will delve into the definition, properties, and examples of cyclic codes, highlighting their significance in coding theory.

9.2.1   Definition of Cyclic Codes

Cyclic codes are linear block codes that possess the cyclic shift property. Mathematically, a cyclic code of length n over a finite field $\mathbb{F}_q$ is defined by the polynomial $g(x)$ of degree $n - k$, where k is the dimension of the code. The codewords are obtained by taking cyclic shifts of the code polynomial.

$$C : \mathbb{F}_q^k \to \mathbb{F}_q^n$$

9.2.2   Generator Polynomial

The key element in cyclic codes is the generator polynomial $(g(x))$. The cyclic shift property implies that if $c(x)$ is a codeword, then $xc(x)$ and $x^2c(x)$ are also codewords. The generator polynomial is often represented in its polynomial form:

$$g(x) = (x - \alpha_1)(x - \alpha_2)\ldots(x - \alpha_{n-k})$$

where $\alpha_1, \alpha_2, \ldots, \alpha_{n-k}$ are the roots of $g(x)$.

9.2.3   Example: Binary Cyclic Code

Consider a binary cyclic code with the generator polynomial $g(x) = x^3 + x + 1$. The roots of this polynomial, denoted as $\alpha_1, \alpha_2, \alpha_3$, are the primitive elements of the field $\mathbb{F}_{2^3}$.

9.2.4   Encoding in Cyclic Codes

The encoding process in cyclic codes involves dividing the message polynomial $m(x)$ by the generator polynomial $g(x)$ to obtain the codeword polynomial $c(x)$. This process can be efficiently implemented using polynomial division or algebraic methods like the Euclidean algorithm.

$$m(x) \xrightarrow{g(x)} c(x)$$

9.2.5 Example: Encoding in Binary Cyclic Code

Let $m(x) = x^2 + x$, and $g(x) = x^3 + x + 1$ be the generator polynomial. The codeword $c(x)$ is obtained by dividing $m(x)$ by $g(x)$:

$$x^2 + x \xrightarrow{x^3+x+1} x^2 + 1$$

The codeword $c(x) = x^2 + 1$ is the transmitted sequence.

9.2.6 Decoding Process

Decoding in cyclic codes involves finding the remainder when the received polynomial $r(x)$ is divided by the generator polynomial $g(x)$. This remainder is the error locator polynomial, and the errors can be corrected based on its roots.

$$r(x) \xrightarrow{g(x)} \text{remainder}$$

9.2.7 Example: Decoding in Binary Cyclic Code

Let $r(x) = x^2 + x + \epsilon(x)$, where $\epsilon(x)$ represents the error polynomial. The decoding process involves dividing $r(x)$ by the generator polynomial $g(x)$ to obtain the remainder. The roots of the remainder polynomial indicate the positions of errors.

$$x^2 + x + \epsilon(x) \xrightarrow{x^3+x+1} \text{remainder}$$

If the remainder is zero, no errors are detected. Otherwise, error correction is performed based on the roots of the remainder.

9.2.8 Cyclic Redundancy Check (CRC)

Cyclic codes are widely used in error detection, and one common application is the Cyclic Redundancy Check (CRC). CRC codes are designed to detect a broad range of errors with a polynomial divisor.

9.2.9   Example: CRC in Network Communication

In network communication, CRC codes are employed to check for errors in transmitted data frames. The polynomial divisor is predetermined, and the remainder obtained during the CRC check helps identify potential errors.

9.2.10   Applications in Digital Storage

Cyclic codes find applications in digital storage systems, where they contribute to error correction and detection mechanisms. Examples include CDs, DVDs, and flash drives, where data integrity is crucial.

9.2.11   Example: Error Correction in DVDs

DVDs often use cyclic codes, such as Reed-Solomon codes, for error correction. These codes can correct burst errors commonly encountered in optical storage.

9.2.12   Applications in Telecommunication

Cyclic codes play a vital role in telecommunication systems, ensuring reliable data transmission over communication channels. The cyclic shift property aids in efficient encoding and decoding processes.

9.2.13   Example: Mobile Communication

In mobile communication systems, cyclic codes are utilized to enhance the reliability of transmitted signals. Error correction mechanisms based on cyclic codes contribute to improved call quality.

9.2.14   Research Frontiers

Ongoing research in cyclic codes explores advanced topics such as the design of codes with specific properties, efficient encoding and decoding algorithms, and their application in emerging technologies like IoT and 5G communication.

9.2.15 Example: Concatenated Codes

Concatenated codes, combining cyclic codes with other types of codes, represent a research frontier. These codes aim to achieve superior error correction capabilities and find application in modern communication systems.

9.3 Error Detection and Correction

Error detection and correction are critical aspects of coding theory, ensuring the reliability of transmitted information in the presence of channel noise or interference. This section explores fundamental techniques for detecting and correcting errors in encoded messages.

9.3.1 Introduction to Error Detection

In digital communication systems, errors can occur during the transmission of information due to various factors such as channel noise, interference, or signal degradation. Error detection is the process of identifying the presence of errors in the received data.

9.3.2 Parity Check

One of the simplest methods for error detection is the parity check. In a binary code, a parity bit is added to each codeword. The parity bit is chosen to make the total number of 1s in the codeword (including the parity bit) even or odd, depending on the chosen parity scheme.

Example: Even Parity

For even parity, if the original message has an odd number of 1s, an additional 1 is added as the parity bit to make the total number of 1s even.

$$\text{Original Message: } 1010101 \quad \Rightarrow \quad \text{Codeword: } 10101011$$

If an error occurs during transmission, the parity check can reveal the presence of an error.

9.3.3   Hamming Code for Error Correction

While parity checks can detect errors, they lack the ability to correct them. Hamming codes, on the other hand, are capable of both error detection and correction. The (7, 4) Hamming code, for example, can correct single-bit errors.

Example: (7, 4) Hamming Code

Consider the binary message $M = 1101$. The Hamming code adds three parity bits to form the codeword C. The positions of the parity bits are determined by powers of 2.

Position	1	2	3	4	5	6	7
Bit	$P1$	$P2$	$M1$	$P4$	$M2$	$M3$	$M4$

The parity bits are calculated based on the positions they cover. For example, $P1$ covers positions 1, 3, 5, 7, so its value is set to ensure even parity for these positions.

$$P1 = M1 \oplus M2 \oplus M4$$
$$P2 = M1 \oplus M3 \oplus M4$$
$$P4 = M2 \oplus M3 \oplus M4$$

The resulting codeword is $C = 1011010$. If a single-bit error occurs during transmission, the parity checks can identify and correct the error.

9.3.4   BCH Codes for Multiple Error Correction

BCH (Bose-Chaudhuri-Hocquenghem) codes are an extension of Hamming codes and can correct multiple errors. These codes are widely used in various communication systems.

Example: (15, 7) BCH Code

Consider a (15, 7) BCH code. The binary message M is encoded into a codeword C using the BCH encoding algorithm. If multiple errors occur during transmission, the BCH decoding algorithm can correct them, making BCH codes robust against errors.

9.3.5 Reed-Solomon Codes

Reed-Solomon codes are another powerful class of error-correcting codes widely used in digital communication and storage systems. These codes can correct both random and burst errors.

Example: Reed-Solomon Code

In a Reed-Solomon code, the message is represented as a polynomial. The encoding process involves evaluating the polynomial at specific points to obtain the codeword. The decoding process uses interpolation to correct errors.

9.3.6 Turbo Codes and LDPC Codes

Turbo codes and LDPC (Low-Density Parity-Check) codes represent modern error correction techniques used in telecommunications. These codes have excellent error-correcting capabilities and find applications in 4G and 5G communication systems.

Example: Turbo Code

A turbo code employs parallel concatenated convolutional codes and iterative decoding. The iterative process refines the estimated information bits, improving the overall decoding performance.

9.3.7 Introduction to Error Detection and Correction in Modular Arithmetic

In the context of modular arithmetic, error detection and correction algorithms need to consider the arithmetic operations performed in a finite field. Techniques such as modular addition, subtraction, multiplication, and division play a crucial role in designing effective error correction codes.

Example: Modular Addition

Consider modular addition in a finite field $\mathbb{F}_p$. If an error causes an incorrect sum during addition, the result may fall outside the valid range of the finite field. Error detection algorithms can identify such discrepancies.

9.3.8 Trade-offs in Error Correction

While advanced error correction codes provide impressive capabilities, they often come with increased computational complexity and higher bandwidth requirements. Designing error correction systems involves trade-offs between performance and resource utilization.

9.3.9 Quantum Error Correction

In the emerging field of quantum computing, quantum error correction becomes essential due to the susceptibility of quantum bits (qubits) to decoherence and other quantum errors. Techniques like Shor's code and the surface code are designed to address these challenges.

Example: Surface Code

The surface code involves encoding qubits on a two-dimensional lattice. The code uses stabilizer measurements to detect and correct errors. The fault-tolerant nature of the surface code contributes to the reliability of quantum computations.

Chapter 10

Future Directions

10.1 Open Problems

The field of modular arithmetic is dynamic, and numerous open problems beckon researchers to explore new avenues, discover novel solutions, and advance our understanding of this fascinating branch of mathematics. In this section, we highlight some of the intriguing open problems that await exploration.

10.1.1 1. Congruence of Power Sums

One open problem involves the congruence properties of power sums. For which integers a and n do the power sums $1^a + 2^a + \ldots + n^a$ exhibit interesting congruence patterns modulo a prime p?

10.1.2 2. Primitive Roots Existence

The existence of primitive roots for all prime numbers is a classic open problem. Specifically, is there a primitive root for every prime p, and if so, can a general construction method be devised?

10.1.3   3. Distribution of Quadratic Residues

Understanding the distribution of quadratic residues modulo p is a challenging problem. What can be said about the density and distribution of quadratic residues as p varies?

10.1.4   4. Generalized Ramanujan Conjecture

A generalization of Ramanujan's conjecture for partition functions involves understanding the distribution of the number of ways an integer can be expressed as a sum of k squares. What can be said about the asymptotic behavior of these distributions?

10.1.5   5. Modular Forms and Elliptic Curves

The Birch and Swinnerton-Dyer conjecture connects the rank of an elliptic curve over the rational numbers to the order of vanishing of a corresponding modular form at its critical point. Proving this conjecture for a broader class of elliptic curves remains an open problem.

10.1.6   6. Prime Number Races

The study of prime number races involves understanding the distribution of prime values of polynomials. Can one prove the existence of infinitely many prime values for certain families of polynomials?

10.1.7   7. Arithmetic Dynamics

In the realm of arithmetic dynamics, the existence of periodic points for rational functions over finite fields is an open problem. What can be said about the behavior of such functions in finite fields?

10.1.8 8. Generalizations of Fermat's Last Theorem

Exploring generalizations of Fermat's Last Theorem in various arithmetic settings remains an open problem. What conditions ensure the solvability or unsolvability of certain Diophantine equations?

10.1.9 9. Algebraic Independence

The algebraic independence of certain modular forms and their values at algebraic points is a deep open problem. Can one establish the algebraic independence of values of certain modular functions?

10.1.10 10. Explicit Construction of Elliptic Curves

Providing an explicit construction of elliptic curves with prescribed properties is an open problem. Can one find a general method for constructing elliptic curves with specific invariants?

10.1.11 11. Uniform Distribution of Sequences

The uniform distribution of sequences in modular arithmetic is a topic of ongoing research. For which sequences do the terms exhibit a uniform distribution modulo p as p varies?

10.1.12 12. Modular Forms and L-functions

The relationship between modular forms and L-functions is a central theme in number theory. Understanding the analytic properties of these L-functions and establishing further connections with modular forms is an open problem.

10.1.13 13. Solvability of Diophantine Equations

The solvability of Diophantine equations in specific polynomial rings is an open problem. What can be said about the existence of solutions and the complexity of solving these equations over finite fields?

10.1.14   14. Modular Abelian Varieties

The study of modular abelian varieties involves exploring their arithmetic properties. What is the structure of the endomorphism ring of a modular abelian variety, and how does it relate to its geometry?

10.1.15   15.   Computational Aspects of Modular Arithmetic

The efficient computation of modular arithmetic operations for large integers is a practical open problem. Can algorithms be devised that optimize modular addition, subtraction, multiplication, and exponentiation for cryptographic applications?

10.2   Trends in Research

The landscape of modular arithmetic is continually evolving, with researchers exploring new directions, innovative methodologies, and interdisciplinary connections. This section delves into some of the current trends in research within the realm of modular arithmetic.

10.2.1   1. Applications in Cryptography

One prominent trend is the ever-growing role of modular arithmetic in cryptography. Modular operations form the foundation of public-key cryptography algorithms such as RSA. The security of these systems relies on the difficulty of certain modular arithmetic problems, such as factoring large semiprime numbers.

10.2.2   2. Arithmetic Dynamics

Arithmetic dynamics, a field at the intersection of number theory and dynamical systems, has gained momentum. Researchers explore the behavior of sequences generated by iterating arithmetic operations in modular arithmetic. This area opens avenues for studying long-term behavior and periodicity in these sequences.

10.2.3   3. Computational Aspects

Advancements in computational methods for modular arithmetic have garnered attention. Researchers focus on developing efficient algorithms for modular exponentiation, modular inversion, and other fundamental operations. These advancements have implications for applications in cryptography, coding theory, and numerical analysis.

10.2.4   4. Modular Forms and L-functions

The study of modular forms and their associated L-functions remains a vibrant area of research. Connections between modular forms, elliptic curves, and L-functions have profound implications for understanding the distribution of prime numbers and solving Diophantine equations.

10.2.5   5. Quantum Modular Forms

Recent trends include the exploration of quantum modular forms, bringing together concepts from quantum field theory and modular forms. These forms arise in the study of certain aspects of string theory and have connections to mock modular forms.

10.2.6   6. Algebraic Structures in Modular Arithmetic

Researchers investigate algebraic structures within modular arithmetic, exploring properties of modular rings, fields, and polynomials. Understanding the structure of these algebraic objects contributes to a deeper comprehension of modular arithmetic itself.

10.2.7   7. Modular Abelian Varieties

The study of modular abelian varieties has gained prominence. Researchers examine the arithmetic properties of these varieties, connections with Galois representations, and their role in the Langlands program.

10.2.8   8. Explicit Formulae in Modular Arithmetic

Developing explicit formulae for various arithmetic functions in modular arithmetic is an active area. Explicit formulae provide insights into the distribution of values and play a crucial role in analytic number theory.

10.2.9   9. Coding Theory and Modular Arithmetic

Connections between coding theory and modular arithmetic continue to be explored. Researchers investigate the use of modular codes for error detection and correction, particularly in the context of arithmetic codes and cyclic codes.

10.2.10   10. Random Matrix Theory and Modular Arithmetic

Random matrix theory has found applications in studying the statistics of zeros of L-functions and other arithmetic objects. Researchers explore connections between random matrix theory and modular forms, uncovering deep relationships.

10.2.11   11. Modular Arithmetic in Quantum Computing

With the rise of quantum computing, the study of modular arithmetic in the quantum realm is an emerging trend. Quantum algorithms for modular exponentiation and factorization pose new challenges and opportunities for exploration.

10.2.12   12. Homotopy Theory and Modular Forms

Homotopy theory enters the scene, providing a novel perspective on modular forms. Researchers investigate homotopy invariants associated with modular forms, opening new avenues for understanding the topology of moduli spaces.

10.2.13   13. Arithmetic Geometry and Modular Curves

Arithmetic geometry techniques play a crucial role in the study of modular curves. Researchers explore the arithmetic properties of modular curves and their applications in solving Diophantine equations.

10.2.14   14. Modular Arithmetic in Machine Learning

The intersection of modular arithmetic and machine learning is a burgeoning area. Researchers explore the potential applications of modular arithmetic in developing robust algorithms for certain types of machine learning problems.

10.2.15   15. Educational Initiatives in Modular Arithmetic

A trend in recent years involves educational initiatives to make modular arithmetic more accessible. Researchers and educators collaborate to develop interactive tools, online courses, and outreach programs aimed at fostering a deeper understanding of modular arithmetic among students.

Chapter 11

Conclusion

11.1 Summary

As we conclude our exploration of modular arithmetic, it is valuable to summarize the key concepts, theorems, and applications discussed in this comprehensive journey through the fascinating world of modular arithmetic.

11.1.1 1. Basics of Modular Arithmetic

At the heart of modular arithmetic lies the concept of congruence. For integers a, b, and a positive integer n, we say that a is congruent to b modulo n if $(a - b)$ is divisible by n. This relationship is denoted as $a \equiv b \pmod{n}$. The set of integers modulo n forms a residue class, and arithmetic operations in this set follow the rules of modular arithmetic.

11.1.2 2. Modular Addition and Subtraction

Modular addition and subtraction involve finding the sum or difference of two numbers modulo n. For example, in $\mathbb{Z}_7$, the sum of 5 and 4 is $5 + 4 \equiv 2 \pmod{7}$, as it cycles back to 2 after reaching 7.

11.1.3   3. Modular Multiplication

In modular multiplication, we find the product of two numbers modulo n. For instance, in $\mathbb{Z}_{11}$, the product of 6 and 8 is $6 \times 8 \equiv 10 \pmod{11}$.

11.1.4   4. Modular Exponentiation

Modular exponentiation is a crucial operation, especially in the context of cryptography. For integers a and k, computing $a^k \pmod{n}$ efficiently plays a vital role in various algorithms.

Example: Modular Exponentiation

Let's compute $3^{13} \pmod{7}$. Using the square-and-multiply algorithm, we get $3^{13} \equiv 6 \pmod{7}$.

11.1.5   5. Congruence Classes and Residue Systems

Understanding congruence classes and residue systems allows us to group integers with the same remainder when divided by n. These classes form the foundation for modular arithmetic operations.

11.1.6   6. Chinese Remainder Theorem

The Chinese Remainder Theorem provides a method to solve systems of simultaneous congruences efficiently. For example, given $x \equiv 2 \pmod{3}$ and $x \equiv 3 \pmod{5}$, the Chinese Remainder Theorem yields $x \equiv 8 \pmod{15}$.

11.1.7   7. Applications in Cryptography

Modular arithmetic plays a central role in cryptographic systems. The RSA algorithm, for instance, relies on the difficulty of factoring large semiprime numbers, which is a problem closely related to modular arithmetic.

11.1.8 8. Error Detection and Correction

In the realm of coding theory, modular arithmetic finds applications in error detection and correction codes. Techniques such as Hamming codes and Reed-Solomon codes leverage the properties of modular arithmetic to ensure reliable data transmission.

11.1.9 9. Number Theory and Diophantine Equations

Modular arithmetic is deeply connected to number theory, particularly in the study of Diophantine equations. Solving equations like $ax + by = c$ modulo n involves techniques from modular arithmetic.

11.1.10 10. Advanced Topics: Fermat's Little Theorem

Fermat's Little Theorem is a powerful result stating that if p is a prime number and a is an integer not divisible by p, then $a^{p-1} \equiv 1 \pmod{p}$. This theorem has applications in primality testing and modular exponentiation algorithms.

11.1.11 11. Future Directions and Open Problems

Our journey concludes with a glimpse into the future directions and open problems in modular arithmetic. These include exploring advanced topics like quantum modular forms, arithmetic dynamics, and the applications of modular arithmetic in emerging fields such as quantum computing and machine learning.

11.2 Closing Remarks

As we approach the conclusion of our journey through the intricate realm of modular arithmetic, it is fitting to reflect on the significance of the concepts explored, the applications uncovered, and the beauty inherent in the simplicity of modular arithmetic.

11.2.1   1. The Elegance of Modular Arithmetic

At its core, modular arithmetic unveils a remarkable elegance. The notion of congruence, encapsulated in $a \equiv b \pmod{n}$, transcends complexity with its intuitive simplicity. This elegance resonates through the entire landscape of modular arithmetic, from basic operations to advanced theorems.

11.2.2   2. Bridge Between Number Theory and Algebra

Modular arithmetic serves as a powerful bridge between number theory and algebra. The study of congruences, residue systems, and modular forms enriches our understanding of both algebraic structures and number-theoretic phenomena.

11.2.3   3. Applications in Cryptography

The practical applications of modular arithmetic are vast and impactful. Cryptographic systems, such as RSA, harness the difficulty of modular arithmetic problems for secure communication. For example, the encryption process involves modular exponentiation, ensuring the confidentiality of transmitted information.

Example: RSA Encryption

Let $p = 17$, $q = 19$, and $e = 5$ be public keys. If a user wants to encrypt a message $M = 8$, the ciphertext C is calculated as $C \equiv 8^5 \pmod{323}$.

11.2.4   4. Error Detection and Correction in Coding Theory

In coding theory, modular arithmetic finds practical use in the creation of robust error-detecting and error-correcting codes. The systematic structure of these codes, often based on modular operations, ensures reliable data transmission even in the presence of errors.

Example: Hamming Code

Consider a Hamming code with three parity bits added to a 4-bit message. The code is designed to detect and correct single-bit errors. The positions of the parity bits are determined using modular arithmetic.

11.2.5   5. Insights into Diophantine Equations

The study of Diophantine equations, equations that involve integer solutions, is enriched by modular arithmetic techniques. The solutions to equations like $ax + by = c$ modulo n provide insights into the existence and nature of integer solutions.

Example: Linear Diophantine Equation

Solving the equation $21x + 15y = 3$ modulo 6 allows us to find solutions in integers. The modular approach provides a systematic method to explore the solution space.

11.2.6   6. Advanced Concepts: Fermat's Little Theorem

Fermat's Little Theorem stands as a testament to the deep connections between number theory and modular arithmetic. The theorem's elegance and simplicity mask its profound implications, particularly in the field of modular exponentiation.

Example: Primality Testing

Fermat's Little Theorem forms the basis for certain primality testing algorithms. If $a^{p-1} \not\equiv 1 \pmod{p}$, then p is definitely not prime, offering a probabilistic primality testing method.

11.2.7   7. Future Horizons and Open Questions

As we stand at the cusp of the conclusion, it's essential to acknowledge that modular arithmetic is far from exhausted in its possibilities. The field continues to evolve, presenting open questions and beckoning future explorations into uncharted territories.

Open Question: Riemann Hypothesis Connection

One intriguing open question involves the potential connections between modular forms and the Riemann Hypothesis. Exploring these connections could deepen our understanding of both modular arithmetic and the distribution of prime numbers.

11.2.8   8. A Call to Exploration

In closing, modular arithmetic extends an invitation to all mathematical explorers. Whether delving into the subtleties of modular forms, unraveling the mysteries of advanced theorems, or applying modular arithmetic in diverse fields, there is always more to discover.

11.2.9   9. Gratitude and Acknowledgments

Before we bid farewell to the world of modular arithmetic, gratitude is extended to the mathematical community, educators, and learners. The collaborative spirit in the pursuit of mathematical knowledge is the driving force behind the continued vibrancy of modular arithmetic.

11.2.10   10. A Continuum of Learning

As we conclude this book, it is our hope that the exploration of modular arithmetic serves as a stepping stone for further learning. The journey is not confined to these pages but extends into the vast expanse of mathematical inquiry.

11.2.11   11. Closing the Chapter

In the closing chapter of our odyssey, let us remember that modular arithmetic, with its simplicity and depth, is a chapter in the grand narrative of mathematics. Closing this chapter only opens the door to new mathematical adventures.

11.2.12   12. Farewell and New Beginnings

With gratitude for the insights gained, challenges faced, and discoveries made, we bid farewell to the captivating world of modular arithmetic. Yet, in every farewell, there is the promise of new beginnings, new questions, and new mathematical horizons to explore.

11.2.13   13. Final Thoughts

In our final thoughts, let us carry the essence of modular arithmetic—a blend of elegance, utility, and perpetual curiosity—into our mathematical endeavors. As we part ways with this exploration, let it be a catalyst for future mathematical journeys.

11.2.14   14. Closing the Book

As we close the book on modular arithmetic, each theorem, example, and concept within these pages becomes a beacon in the vast mathematical landscape. The book may close, but the ideas within reverberate, inspiring continued exploration and discovery.

11.2.15   15. An Invitation to Return

To the reader, we extend an invitation to return to the world of modular arithmetic. The concepts explored here are timeless, and there is always room for further investigation, contemplation, and appreciation. The journey may conclude, but the door remains open for those eager to delve once more into the beauty of modular arithmetic.